LONG-TERM P

HOLT, RINEHART AND WINSTON

A Harcourt Classroom Education Company

Austin • New York • Orlando • Atlanta • San Francisco • Boston • Dallas • Toronto • London

To the Teacher

Long-Term Projects contains blackline masters that complement regular classroom use of *Geometry*. These projects are especially helpful in accommodating students of varying interests, learning styles, and ability levels. There is one four-page Long-Term Project for each chapter of *Geometry*. These projects engage students in activities that encompass more than one class period. Projects often require data collection or research to be done outside the classroom. Many of the projects are appropriate for group work.

Copyright © by Holt, Rinehart and Winston

All rights reserved. No part of this publication may be reproduced or transmitted in any form or by any means, electronic or mechanical, including photocopy, recording, or any information storage and retrieval system, without permission in writing from the publisher.

Teachers using GEOMETRY may photocopy complete pages in sufficient quantities for classroom use only and not for resale.

Photo Credit
Front Cover: Dale Sanders/Masterfile.

Printed in the United States of America

ISBN 0-03-054323-1

5 6 7 066 03

Table of Contents

Chapter 1	The Streets of San Francisco	**1**
Chapter 2	License to DRIVE!	**5**
Chapter 3	The Newest Mall in America	**9**
Chapter 4	Terrific Tiles	**13**
Chapter 5	FIRE! Out of Control	**17**
Chapter 6	The "Rice" Report	**21**
Chapter 7	Swimming, Swimming in Our Swimming Pool	**25**
Chapter 8	Similar People	**29**
Chapter 9	Excursions into the History of Pi	**33**
Chapter 10	Congratulations! You're Officially a Mathematician	**37**
Chapter 11	A Fractal Cut-Up	**41**
Chapter 12	"Venn" Ever	**45**
Answers		**49**

NAME _____ CLASS _____ DATE _____

Long-Term Project
The Streets of San Francisco, Chapter 1, page 1

Golden Gate Bridge, cable cars, Union Square—there's lots to see in San Francisco. Historically, San Francisco is well known for the great earthquake and fire of 1906 and another giant quake in 1989.

This map of San Francisco shows some famous tourist stops. To complete this project, you should make at least three photocopies of this map so you can draw on the map and make measurements. Or, if you like, obtain your own larger map of San Francisco.

Radio stations sometimes sponsor special promotions where you can win by locating a hidden treasure. Suppose that a San Francisco radio station, KFUN, hides a treasure. Each day, the station broadcasts clues about the treasure's location. In this project, you will hunt for hidden treasure and then design your own hidden treasure contest!

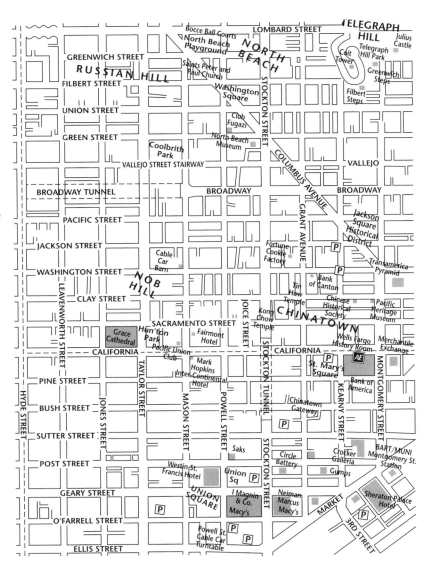

Part I: Use the clues to find the hidden treasure. Complete the questions following the clues. The questions will help you to think about writing your own clues in Part II.

CLUE 1: "This is radio station KFUN. The first clue for the FUNtastic Treasure Hunt is: Get out your maps! Label Coit Tower as point A, the Transamerica Pyramid as point B (place B approximately in the center of the marking), and the Fortune Cookie Factory as point C. Connect the points to form a triangle."

1. Describe $\triangle ABC$. Include both the angle measures and the side lengths in your description.

_____

_____

Geometry — Long-Term Project

NAME _____ CLASS _____ DATE _____

Long-Term Project
The Streets of San Francisco, Chapter 1, page 2

CLUE 2: "This is radio station KFUN. The second clue for the FUNtastic Treasure Hunt is: Get out your maps! Draw the angle bisector of ∠ABC. Label the point of intersection of the angle bisector with $\overline{AC}$ as D."

2. What is the approximate location of D? _____

3. Is $\overline{BD}$ a median of ∠ABC? Explain. _____

   _____

4. Is $\overline{BD}$ an altitude of ∠ABC? Explain. _____

   _____

CLUE 3: "This is radio station KFUN. The third clue for the FUNtastic Treasure Hunt is: Get out your maps! Reflect △ABC across a line down the center of Stockton Street. Label the vertices A', B', C'. Be sure to mark the location D'."

5. Describe the approximate locations of the vertices of △A'B'C'. _____

   _____

CLUE 4: "This is radio station KFUN. The fourth clue for the FUNtastic Treasure Hunt is: Get out your maps! Reflect △A'B'C' across a line down the center of Clay Street. Label the vertices A'', B'', C''. Be sure to mark the location of D''."

6. Describe the approximate locations of the vertices of △A''B''C''. _____

   _____

7. Is there a single translation or rotation that would transform △ABC to

   △A''B''C''? Explain. _____

   _____

CLUE 5: "This is radio station KFUN. The fifth clue for the FUNtastic Treasure Hunt is: Get out your maps! Locate D and D''. Draw a triangle such that all sides are congruent. One vertex will be D, the second vertex will be D'', and the third vertex will be located near Pacific and Jones. Label the third vertex X."

8. Describe how you drew a triangle with three congruent sides using the two given vertices.

   _____

   _____

2 Long-Term Project **Geometry**

Long-Term Project

The Streets of San Francisco, Chapter 1, page 3

CLUE 6: "This is radio station KFUN. The sixth clue for the FUNtastic Treasure Hunt is: Get out your maps! Draw the perpendicular bisectors of each side of $\triangle DD''X$. Label the intersection point of the perpendicular bisectors as E."

9. Describe the approximate location of E. _____

CLUE 7: "This is radio station KFUN. The seventh clue for the FUNtastic Treasure Hunt is: Get out your maps! From E proceed towards Hyde Street. Walk about a block. Look to your right. You should see a famous location labeled on the map."

10. What is the famous location? _____

Before proceeding to the next clue, solve the following two sets of exercises. These exercises will help you to locate compass headings and find distances on the map.

In some cities, the streets run exactly north and south. San Francisco streets are not quite that easy. Locate Hyde Street. As it runs toward the top of the map, its compass heading is approximately 350. Find the compass headings to various locations using the given information. Use a protractor to check your answers.

11. The sum of the compass heading, its complement, and its supplement from the Cable Car Barn to the North Beach Museum is 250°. What is the compass heading from the Cable Car Barn to the North Beach Museum?

_____

12. Three times the compass heading from the Bank of Canton to the Wells Fargo History Room plus 40° is 400°. What is the compass heading from the Bank of Canton to the Wells Fargo History Room?

_____

Use a ruler to estimate each distance in miles. The distance from Hyde Street to Grant Avenue is about one-half mile.

13. From Coit Tower to the Transamerica Pyramid _____

14. From the Transamerica Pyramid to the Fortune Cookie Factory _____

CLUE 8: "This is radio station KFUN. The eighth clue for the FUNtastic Treasure Hunt is: Get out your maps! From the famous location you found in CLUE 7, proceed at a heading of about 170 for approximately 0.14 miles. You should be at another famous location."

15. What is the famous location? _____

NAME _____ CLASS _____ DATE _____

Long-Term Project

The Streets of San Francisco, Chapter 1, page 4

CLUE 9: "This is radio station KFUN. The ninth and final clue for the FUNtastic Treasure Hunt is: Get out your maps! From the famous location you found in CLUE 8, proceed on California Street towards Hyde Street. The treasure is located on California Street such that it is three times as far from Montgomery Street as from Hyde Street. The treasure is inside a large tourist attraction."

16. Where is the treasure located? _____

Before designing your own treasure hunt, answer these questions.

17. Your friend uses a map that is twice as large as your San Francisco map. How does the size of the map affect the angle and distance measures?

_____

_____

18. How did you find distance, in miles, on the map?

_____

_____

19. How did you use the heading of 350 for Hyde Street to find other compass headings?

_____

_____

Part II: Design a treasure hunt. To complete this part of the project, follow these steps.

   Step 1: Obtain a map for a city of your choice. Be sure the map has some tourist attractions labeled. Check your local library or the internet for city maps.

   Step 2: Write at least six clues similar to the ones in **Part I**.

   Step 3: Make photocopies of your map and show how the treasure can be located using the clues.

   Step 4: Give another student or group of students a copy of your city map and your clues. Have the student or group find the hidden treasure.

   Step 5: Check the student's or group's solution to your treasure hunt.

NAME _____ CLASS _____ DATE _____

Long-Term Project
License to DRIVE!, Chapter 2, page 1

Since driving a motor vehicle is a privilege rather than a right, rules govern when and under what conditions a person may get a license. The table below shows the age requirements for applying for a regular license and a learner's permit. In this project, you will follow a student through the logical process of obtaining a license and write a step-by-step procedure one can follow to get one.

Age Requirements for Applying for a Regular License or a Learner's Permit					
State	**Age (reg. license)**	**Age (learner's permit)**	**State**	**Age (reg. license)**	**Age (learner's permit)**
Alabama	16	15	Montana	15	14y, 6mon
Alaska	16	14	Nebraska	17	15
Arizona	16	15y, 7mon	Nevada	16	15y, 6mon
Arkansas	16	14	New Hampshire	18	16
California	17	15	New Jersey	17	16
Colorado	16	15y, 3mon	New Mexico	15	15
Connecticut	16y, 6mon	16	New York	18	16
Delaware	15y, 10mon	14y, 10mon	North Carolina	16y, 6mon	15
Dist. of Col.	16	16	North Dakota	16	14
Florida	18	15	Ohio	17	15y, 6mon
Georgia	18	15	Oklahoma	16	15y, 6mon
Hawaii	15	15	Oregon	16	15
Idaho	15	15	Pennsylvania	18	16
Illinois	17	15	Rhode Island	17	16
Indiana	18	15	South Carolina	16y, 3mon	15
Iowa	17	14	South Dakota	14	14
Kansas	16	14	Tennessee	16	15
Kentucky	16y, 6mon	16	Texas	16	15
Louisiana	17	15	Utah	16	16
Maine	16	15	Vermont	16	15
Maryland	17y, 7mon	15y, 9mon	Virginia	16	15
Massachusetts	18	16	Washington	16	15
Michigan	17	14y, 9mon	West Virginia	16	15
Minnesota	17	15	Wisconsin	16	15y, 6mon
Mississippi	16	15	Wyoming	16	15
Missouri	16	15y, 6mon			

Part I: Follow a student. Complete the questions to help you understand the process involved in obtaining a regular license.

1. Kristin is a student at Belgrade High School in Montana. Use the information in the table to write a true conditional statement about the age for applying for a regular license in Montana.

_____

Geometry   Long-Term Project

Long-Term Project
License to DRIVE!, Chapter 2, page 2

2. Write the converse of your conditional statement. _____
   _____

3. Is the converse you wrote in Exercise 2 true? Explain. _____
   _____

4. On a separate sheet, write a conditional and its converse for your state using either the age requirement for a learner's permit or a regular license.

In the table at right, some states have been classified according to their age requirements for applying for a regular license and a learner's permit. (The table on page 1 was used to classify the states.)

Group 1	Group 2
Alaska	Dist. of Col.
Arkansas	Hawaii
Kansas	Idaho
North Dakota	New Mexico
	South Dakota
	Utah

5. Write a criterion for telling if a state is in Group 1. _____
   _____

6. Write a criterion for telling if a state is in Group 2. _____
   _____

7. On a separate sheet, write definitions to use for classifying the rest of the states based upon the age requirements for applying for a regular license and a learner's permit. Copy the Group 1 and Group 2 table. Add columns for the groups you have defined, and fill in the rest of the states.

Before you apply for a regular license, it is a good idea to take a driver education course. Some public schools offer this course to their students.

8. On a separate sheet, draw an Euler diagram which conveys the following information:

   If a student is in the ninth grade at Belgrade High School, then the student can take a driver education class.

   Kristin is in the ninth grade at Belgrade High School.

9. What conclusion can you draw about Kristin from the information in Exercise 8?
   _____

Long-Term Project
License to DRIVE!, Chapter 2, page 3

10. On the first day of driver education class, Kristin's instructor wrote these six statements on the chalkboard. The instructor also stated that you must meet the age requirements for the learner's permit and the regular license before you can take the tests.

    a. If you study to take the written test to obtain a learner's permit, then you will pass the written test for the learner's permit.

    b. If you are ready to take the driving test for a regular license, then you will pass the test for a regular license.

    c. If you are in the ninth grade, then you can take a driver education class at school.

    d. If you pass the written test for the learner's permit, then you can practice driving with a licensed driver.

    e. If you take a driver education class at school, then you can study to take the written test to obtain a learner's permit.

    f. If you practice driving with a licensed driver, then you will be ready to take the driving test for a regular license.

On a separate sheet, write the six statements in order to form a logical chain. State the overall conclusion from the beginning to the end of the chain.

In Montana, you must take both a written and a driving test in order to obtain a regular license. During the driving test, you must be able to parallel park your car. This diagram shows one important step in parking your car. Notice the line drawn through the center of the car and the line drawn parallel to the curb. Two angles are formed, ∠1 and ∠4.

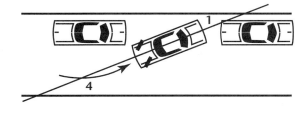

11. Kristin is in her room studying for the test for her learner's permit when her brother, Scott, finds her. Scott tells her that there will be a question on the test for her learner's permit about the relationship of ∠1 and ∠4. Kristin guesses that m∠1 = m∠4. On a separate sheet, write a proof for Kristin.

    Given: m∠1 = m∠3
    m∠4 = m∠5
    m∠2 + m∠4 + m∠5 = 180°

    Prove: m∠1 = m∠4

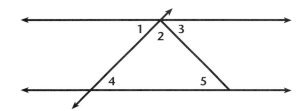

NAME _____ CLASS _____ DATE _____

Long-Term Project
License to DRIVE!, Chapter 2, page 4

The big day has arrived! Kristin has passed her driver education class and is ready to take the written test for her learner's permit. The only problem is that Scott has to drive her to the Court House to take the test. He says that he will not drive her unless she can solve the following problem.

12. Draw a diagram showing the distance between each location using these clues.

   a. The distance from Kristin and Scott's house to the Bridger View Mall is 8 miles.

   b. The distance from Burger John's to the Court House is 12 miles.

   c. The distance from Kristin and Scott's house to the Court House is 15 miles.

   d. Going from West to East, the locations in order are Kristin and Scott's house, Burger John's, the Bridger View Mall, and the Court House.

The Court House is located at the center of four streets that converge as shown in the diagram. Scott insists that Kristin find the missing angle measures if she wants to practice driving his car. Scott tells Kristin that m∠GBN = 35°, m∠MBU = 2m∠GBN, and m∠MBF = 1.5m∠SBT.

Find the following measures.

13. m∠UBS _____

14. m∠MBU _____ **15.** m∠SBT _____ **16.** m∠MBF _____

17. m∠FBG _____ **18.** m∠NBA _____ **19.** m∠TBA _____

Kristin practices driving with her brother Scott. When she meets the age requirement, she takes the driving portion of the test for her regular license. She passes! Now she hopes that she can buy a car!!!

20. To get her regular license, Kristin had to follow a step-by-step procedure. First, she enrolled in a driver education course, then she met an age requirement, then she took a test for a learner's permit, and so on. On a separate sheet, write the steps Kristin followed.

Part II: Write your own procedure. To complete this project, follow these steps. You may want to display your project in a booklet or on a poster.

   Step 1: Research the license requirements in your state. For example, must you take a driver education course? Are there specific skills you must demonstrate when you drive?

   Step 2: Write the steps you would need to follow to obtain a regular license in your state. Include at least five steps.

   Step 3: Use Step 2 to help you write a logic chain similar to the one in Exercise 10. State the overall conclusion from the beginning to the end of the chain.

NAME _____ CLASS _____ DATE _____

Long-Term Project
The Newest Mall in America, Chapter 3, page 1

You have probably heard about the largest shopping mall in the world. It is the West Edmonton Mall in Alberta, Canada. The largest mall in the United States is the Mall of America in Bloomington, MN. Each of these malls has a large indoor amusement park. The amusement park in the Mall of America occupies 7 acres. In this project, you will help design an amusement park to be located in a gigantic mall and then design your own amusement park. The new amusement park will be located at the center of the mall. It will be a regular hexagon measuring 400 feet on each side. ACTION GALORE! will offer excitement for all ages!

Part I: Help to design a new amusement park. Complete the questions in Part I to help you think about how you will design your own amusement park.

1. Describe the rotational symmetries of the amusement park. _____

2. How many axes of symmetry does the amusement park have? _____

3. What is the interior angle sum of the amusement park? _____

4. What is the measure of an interior angle of the amusement park? _____

At the entrance to ACTION GALORE! will be a large flower garden that blooms all year. A diagram of the garden is shown. All four sides of the garden are congruent, but the garden is not a square.

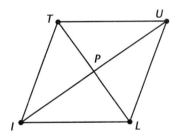

5. What is m∠TPU? _____

6. If m∠PIL = 25°, what is m∠TIP? Explain how you found the angle measure. _____

7. If m∠PIL = 25°, what is m∠UTL? Explain how you found the angle measure. _____

8. Does TL = UI? Explain. _____

9. Suppose the garden is changed so that opposite sides are congruent, but consecutive sides are not. How will your answers to Exercises 5–8 change? _____

Geometry **Long-Term Project** 9

NAME _____ CLASS _____ DATE _____

Long-Term Project

The Newest Mall in America, Chapter 3, page 2

ACTION GALORE! will have a huge Ferris wheel. If you connect the cars with segments, you will have an octagon. To maintain stability, you must be sure that the segments EF and GH are parallel. On a separate sheet, write your own proof, using any form. (Challenge: Write a proof without using this theorem: If two coplanar lines are parallel to the same line, then the two lines are parallel to each other.)

10. Given: $\overline{AB} \parallel \overline{CD}$, $\overline{AB} \parallel \overline{EF}$, and $\overline{CD} \parallel \overline{GH}$
    Prove: $\overline{EF} \parallel \overline{GH}$

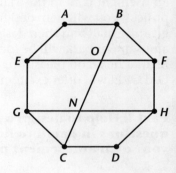

The Ferris wheel will have a number of supports to maintain stability. On a separate sheet, trace the diagram for the Ferris wheel, but do not draw $\overline{BC}$. Draw a segment connecting points B and D. Locate point I such that it is the midpoint of $\overline{BF}$. Locate point J such that it is the midpoint of $\overline{DH}$. Draw $\overline{IJ}$. This type of support system will be used completely around the Ferris wheel. Each side of the octagon measures 44 feet and BD is 150 feet.

Find the measures of the following segments and angles.

11. BI _____ 12. IJ _____ 13. $m\angle BFH$ _____

14. $m\angle FIJ$ _____ 15. $m\angle HJI$ _____ 16. $m\angle FBD$ _____ 17. $m\angle HDB$ _____

The Ferris wheel will be anchored to a rectangular platform. A diagram of the platform is shown.

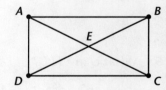

18. If $AE = 53$ ft, how long is AC? _____

19. If $AE = 53$ ft, how long is BD? _____

20. If $m\angle EDC = 32°$, what is $m\angle EAD$? Explain how you found $m\angle EAD$. _____
    _____

21. Estimate the length and width of the platform. Explain your estimation method. _____
    _____
    _____

Long-Term Project

The Newest Mall in America, Chapter 3, page 3

Inside the amusement park, there will be a restaurant that rotates completely every 30 minutes. The restaurant is a regular pentagon labeled PENTA in the diagram. Segments $\overline{BC}$ and $\overline{DF}$ are parallel. In order to build the restaurant correctly, the contractor needs to know the measures of a number of angles.

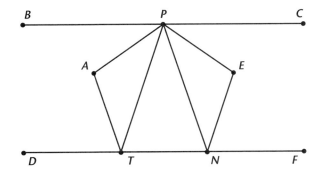

Find the measures of the following angles.

22. m∠PAT _____
23. m∠APE _____
24. m∠BPA _____
25. m∠APT _____
26. m∠ATP _____
27. m∠PTN _____
28. m∠PNT _____
29. m∠TPN _____
30. m∠ENF _____
31. m∠ATD _____
32. m∠ENT _____
33. m∠PEN _____

34. Explain how you found the measures of the angles for the restaurant.

_____

_____

_____

35. What is the measure of each central angle of the restaurant? _____

36. If one complete rotation takes 30 minutes, how many degrees does the restaurant rotate in one minute? _____

37. How long does it take the restaurant to rotate the measure of one central angle? _____

38. The head architect for the amusement park project would like a fountain designed for the front of the restaurant. She wants the fountain to be a regular polygon with an exterior angle that is $\frac{1}{5}$ the measure of an interior angle. What will be the measures of the interior and exterior angles of the fountain? _____

39. How many sides will the fountain have? _____

Geometry · Long-Term Project

NAME _____ CLASS _____ DATE _____

Long-Term Project
The Newest Mall in America, Chapter 3, page 4

No amusement park would be complete without a roller coaster. The architect of the roller coaster for ACTION GALORE! has plotted points on a coordinate grid where the roller coaster ends an ascent or begins a descent.

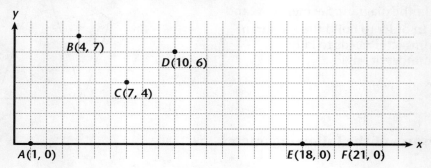

Find the slope of each of the following line segments.

40. $\overline{AB}$ _____ **41.** $\overline{BC}$ _____ **42.** $\overline{CD}$ _____ **43.** $\overline{DE}$ _____ **44.** $\overline{EF}$ _____

45. Explain which segment of the ride you think would be the most scary. Use slope to justify your answer. _____

_____

46. The head architect is considering adding a ride that drops straight down vertically. What would be the slope of this ride? _____

47. On a sheet of graph paper, plot a roller coaster ride of your own design. Find the slope of each segment of the ride. Explain where the scariest parts of your ride are. You may want to include your roller coaster in your amusement park for **Part II**.

Part II: Design an amusement park. To complete this part of the project, follow these steps.

  Step 1: Obtain a poster board or other large paper for displaying your design.

  Step 2: Select an appropriate scale for your design. Decide upon appropriate dimensions for the park. Draw the exterior of the amusement park.

  Step 3: Draw designs for at least two rides and a restaurant. Decide upon appropriate dimensions for the spaces to be occupied by the rides and restaurant.

  Step 4: Draw your rides and restaurant inside the amusement park. If you like, add other features such as gardens, rest areas, and fountains. Label the attractions.

Challenge: Find the area, in acres, of ACTION GALORE!
Find the area of your amusement park. How do the two areas compare to the area of the amusement park located in the Mall of America?

NAME _____ CLASS _____ DATE _____

Long-Term Project
Terrific Tiles, Chapter 4, page 1

Sometimes we fail to notice the details of our surroundings. For example, homes and schools can have floors or walls covered with congruent polygons. Floor tile or wall coverings are often designed using the concept of congruency. Man has had a long history of designing floor coverings. In the Middle Ages, brightly decorated tiles were used both indoors and out. As early as the 4th century, Greeks used arrangements of pebbles to make decorative floor coverings. In this project, you will analyze floor and wall covering designs. Then you will design floor and wall coverings using congruent polygons.

Part I: Look at patterns found in floor and wall coverings. Analyze the figures found in these floor and wall coverings. Complete the questions and make drawings as required.

This design is from a sheet of linoleum that can be purchased to cover a floor. This particular linoleum came in a roll that was fourteen feet wide. The linoleum can be cut to fit the shape of a room and then glued to the subfloor. The diagram shows the pattern that can be seen in a one-foot by one-foot section of the linoleum. This pattern is then repeated over the entire surface of the sheet of linoleum.

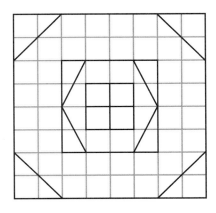

1. Describe the polygons in the design. Include information about congruency, for example, the number of right triangles that appear congruent in the design. _____

   _____

2. The diagram measures two inches by two inches. What are the lengths of the sides of each polygon in the diagram? _____

   _____

3. What will be the approximate dimensions of each polygon on the actual one-foot by one-foot square of this pattern? _____

   _____

4. If you purchased a twelve-foot length of this linoleum, how many one-foot by one-foot sections would there be? Explain how you arrived at your answer. _____

   _____

Geometry   Long-Term Project   **13**

Long-Term Project
Terrific Tiles, Chapter 4, page 2

5. On a separate sheet of paper, draw a four-inch by four-inch square. (Grid paper with one-quarter inch squares will make your drawing easier and more accurate.) Make an accurate diagram of four one-foot by one-foot sections of this tiling pattern joined together. Use your own choice of colors to liven up this tiling design. Save your diagram for use in **Part II** of this project.

You don't have to purchase linoleum in a sheet to cover a floor. Many people, especially those who like to do their own remodeling, are buying self-adhering vinyl tiles that usually measure one-foot by one-foot. These tiles can be cut so that any floor can be easily covered. The diagram shows a one-foot by one-foot square tile.

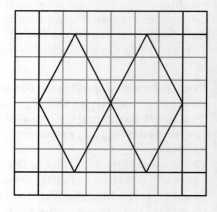

6. Describe the polygons in the design. Include information about the appearance of congruency.

   _____
   _____

7. The diagram measures two inches by two inches. What are the lengths of the sides of each polygon in the diagram? _____
   _____
   _____

8. What will be the approximate dimensions of each polygon on the actual one-foot by one-foot square of this tile? _____
   _____
   _____

9. This particular pattern of tile comes in boxes of 27 tiles. If you purchased two boxes of tile, what could be the dimensions of a room that would use up the entire two boxes of tiles? _____
   _____

10. On a separate sheet of paper, or on grid paper if available, draw a four-inch by four-inch square. Make an accurate diagram of four one-foot by one-foot squares of this tile joined together. Use your choice of colors to liven up this floor tile design. Save your diagram for use in **Part II** of this project.

NAME _____ CLASS _____ DATE _____

Long-Term Project
Terrific Tiles, Chapter 4, page 3

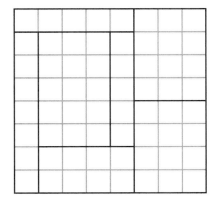

Here is another pattern for a one-foot by one-foot self-adhering tile.

11. Describe the polygons in the design. Include information about apparent congruency. _____

_____

12. The diagram measures two inches by two inches. What are the lengths of the sides of each

polygon in the diagram? _____

_____

13. What will be the approximate dimensions of each polygon on the actual one-foot by one-foot

square tile? _____

_____

14. On a separate sheet of paper or on grid paper, draw a four-inch by four-inch square. Make an accurate diagram of four one-foot by one-foot squares of this tiling pattern joined together. Use your own choice of colors in your diagram, and save it for use in **Part II** of this project.

Congruent polygons are also used in designing wall board that can be used for bathrooms and kitchens. This wall board comes in sheets that measure four feet by eight feet and can be cut to fit various wall sizes. This particular sheet of wall board has four distinct parts: a top border, rows of congruent squares, a center design, and then more rows of congruent squares. The diagram at right shows how the squares, top border, and center design are placed on the entire sheet. Below are diagrams of the top border and the center design.

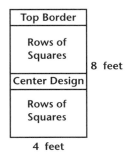

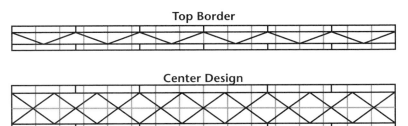

Geometry

Long-Term Project **15**

Long-Term Project

Terrific Tiles, Chapter 4, page 4

15. Describe the polygons in the top border. Include information about perceived congruency.
_____

16. One sheet of wall board is four feet across. The diagram of the top border is drawn such that one inch represents one foot. There are six rectangles across the top border. What are the approximate dimensions of the rectangles on the actual wall board? _____

17. Describe the polygons in the center design. Include information about perceived congruency.
_____

18. What will be the approximate dimensions of the polygons on the actual four-foot by eight-foot sheet of wall board? Include the polygons on both the top border and the center design.
_____
_____

19. In the section following the top border, there are five rows of congruent six-inch by six-inch squares. How many rows of six-inch by six-inch squares are in the section under the center design? _____

20. On a separate sheet of paper or grid paper, draw a four-inch by eight-inch rectangle. Make an accurate diagram of a four-foot by eight-foot sheet of this wall board. Use your own choice of colors in your diagram, and save it for use in **Part II** of this project.

Part II: Make a brochure of floor and wall covering designs. To complete this project, follow these steps.

Step 1: Collect the diagrams you made of floor tile and wall board patterns in Exercises 5, 10, 14, and 20.

Step 2: Draw diagrams of your own designs for two different floor tile patterns. Use congruent polygons in your designs. Show how four one-foot by one-foot sections or tiles fit together. Be sure to add color to your designs.

Step 3: Draw diagrams of your own designs for two different four-foot by eight-foot sheets of wall board. Use congruent polygons in your designs. Be sure to add color to your designs.

Step 4: Make a brochure or booklet of the designs from Steps 1 and 2. Include descriptions of the congruent polygons and their measurements.

Long-Term Project

FIRE! Out of Control, Chapter 5, page 1

In the summer of 1998, wildfires burned more than 150,000 acres of land in Florida. This was the worst wildfire disaster in that state since 1932.

When you hear an amount of land such as 150,000 acres, do you ever wonder how big an area that is? How long would it take to walk or drive around that area?

In this project, you will answer the question: "How long would it take to walk or drive around the perimeter of a 150,000 acre area?" Then you will find your own area statistic and write a problem relating to it.

Part I: Analyze different areas given in acres. Complete the questions and make drawings as required.

Before you determine some possible dimensions for an area of 150,000 acres, let's start with some simpler problems. Suppose that you look in the real estate section of the newspaper and read that a house for sale is located on four acres of land. How big is that?

1. One square mile of land is 640 acres. What portion of a square mile is 4 acres? _____

2. Most people think of the size of a house lot in feet. Find the number of square feet in four acres. Explain how you arrived at your answer. _____

3. Some lots that houses occupy are square. What would be the approximate dimensions, in feet, of a square 4-acre lot? _____

4. What would be the approximate perimeter, in feet, of the lot in Exercise 3? _____

5. How long would it take you to walk around the 4-acre lot? Explain how you arrived at your answer. _____

Geometry — Long-Term Project

Long-Term Project

FIRE! Out of Control, Chapter 5, page 2

6. Suppose that the 4-acre lot was not a square, but a trapezoid. Make a sketch of a trapezoid whose area is 4 acres. Label the dimensions of the lot, in feet.

7. What would be the approximate perimeter, in feet, of the lot you drew in Exercise 6?

8. Suppose that the 4-acre lot was an isosceles right triangle. Make a sketch of an isosceles right triangle whose area is 4 acres. Label the dimensions of the lot, in feet.

9. What would be the approximate perimeter, in feet, of the lot you drew in Exercise 8? _____

10. Compare the perimeters of your square lot, trapezoidal lot, and triangular lot.

Which one would take the longest to walk around? _____

In the Eastern part of Montana, wheat farming is a big operation. Some farmers own the land on which they plant wheat, while others lease large areas of land from the landowner. Suppose that Roger owns several thousand acres of his own land, but also leases 864 acres for planting. Just how big is that?

11. How many squares miles of farm land does Roger lease? _____

12. Some land is plotted in nice, neat shapes. What would be one possible length and width, in miles, of a rectangular-shaped plot covering 864 acres? _____

NAME _____ CLASS _____ DATE _____

Long-Term Project
FIRE! Out of Control, Chapter 5, page 3

13. What would be the approximate perimeter, in miles, of the plot of land in Exercise 12?

   _____

14. How long would it take you to walk around the 864-acre plot of farm land?

   Explain how you arrived at your answer. _____

   _____

   _____

15. Suppose that the 864-acre plot of land was not a rectangle, but closely approximated the shape of a regular hexagon. Make a sketch of a regular hexagon whose area is 864 acres. Label the dimensions of the plot of land, in miles.

16. What would be the approximate perimeter, in miles, of the plot of land in Exercise 15?

   _____

17. Although large plots of land can be shaped as squares or rectangles, the land that is actually planted with a crop, such as wheat, is generally an irregular shape. On a sheet of grid paper, sketch an irregular shape for a plot of farm land whose area is approximately 864 acres. Be sure to include a key defining the area you want each square of the grid to represent.

18. Describe your method for finding the area of the irregular shape you drew in Exercise 17. _____

   _____

   _____

19. Give an estimate for the perimeter of the irregular shape you drew in Exercise 17. _____

   _____

20. How long would it take you to walk around the irregular-shaped 864 acres of farm land? Explain how you arrived at your answer. _____

   _____

Geometry   Long-Term Project   **19**

NAME _____ CLASS _____ DATE _____

Long-Term Project
FIRE! Out of Control, Chapter 5, page 4

Recall that the fire in Florida burned 150,000 acres. How long would it take to walk or drive around the perimeter of a 150,000 acre area? Answer the following questions to help you make an estimate of the time. You will need to decide whether it is more reasonable to estimate the time to walk around the area or the time to drive around the area.

21. How many square miles did the fire destroy? _____

22. If the area destroyed by the fire was approximately circular, what would be the radius, in miles, of the circle? What would be the circumference, in miles, of the circle?

    _____

23. It would be extremely unusual if the area burned by Florida wildfires was shaped as a polygon whose area could be easily calculated. On a sheet of grid paper, draw two possible shapes for the 150,000 acre area. Have the first shape be a combination of several polygons or circles, for example, a square joined to a trapezoid or a right triangle joined to a half-circle. Have the second shape be an irregular shape. Label the dimensions on both drawings and give the area represented by the squares on the grid paper.

24. Give an estimate for the perimeter of each of your shapes in Exercise 23. _____

    _____

    _____

25. Decide whether you want to answer the question using time to walk or time to drive around the 150,000 acre area. Then show your solution to the problem. Be sure to explain how you

    determined what speed to use for walking or driving. _____

    _____

Part II: Write an area problem. To complete this project, follow these steps.

Step 1: Find an interesting statistic involving area, such as the one about the size of the Florida wildfires. Convert the area to acres.

Step 2: Write a problem using the area and the perimeter.

Step 3: Write a detailed solution for your problem. Have another student or group of students solve your problem. Check the solution.

Step 4: Display your problem and solution in a booklet or on a poster or bulletin board.

NAME _____ CLASS _____ DATE _____

Long-Term Project
The "Rice" Report, Chapter 6, page 1

You often hear people make a statement like this: "Look at that box. It's twice as big as the other box." What does that person really mean? Does he or she mean that the box is twice as tall? twice as wide? that it holds twice as much? In this project, you will attempt to answer the question: What does "twice as big" mean? You will construct prisms and compare the amounts of rice that they hold. You will also look at the sum of the areas of the faces of those prisms. Then you will make a display of your prisms and your findings.

For this project you will need poster board or cardboard (cardboard works best), a ruler, a protractor, a measuring cup or cups that will hold at least 1000 mL, and about 1000 mL of uncooked rice.

Part I: Construct prisms and determine how much they hold. Construct prisms as directed and complete the questions. Save all prisms that you construct for use in Part II.

Use a net to construct a rectangular prism. Copy the net onto poster board or cardboard using the given dimensions, in centimeters.

After you draw the net, fold it along the solid lines. Tape the edges together to form a rectangular prism. Leave the top three edges open so that you can lift the top.

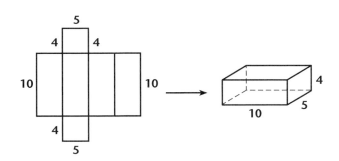

1. Pour rice into the prism until it is completely filled. (If you are using poster board, you will notice that the sides bulge. Try to hold them straight so that you get an accurate measurement of how much the prism holds.) Pour the rice into a measuring cup. Approximately how many mL of rice

   does your prism hold? (This amount is the volume of your prism.) _____

2. Using the formula for area of a rectangle that you learned in Chapter 5, find the area of each face of the prism. What is the sum of the six areas?

   _____

3. Multiply your answer to Exercise 1 by 2. Fill the measuring cup with this amount of rice. Construct a rectangular prism that holds approximately the amount of rice in the measuring cup. On a separate sheet of paper, sketch a net for this prism. Label all dimensions of the prism, in centimeters. If you like, compare your net to the nets of other students.

Geometry   Long-Term Project   **21**

NAME _____ CLASS _____ DATE _____

Long-Term Project
The "Rice" Report, Chapter 6, page 2

4. Using the formula for area of a rectangle, find the area of each face of the prism you constructed in Exercise 3. What is the sum of the six areas? _____

5. If you write a fraction comparing the volumes of the prisms in Exercises 3 and 1, the fraction is $\frac{2}{1}$. Write a fraction in simplest form showing how the sums of the areas of the faces compare (your answers for Exercises 4 and 2). Use the answer for Exercise 4 as the numerator of your fraction. How does this fraction compare to $\frac{2}{1}$? Explain whether you think the prism in Exercise 3 is "twice as big" as the prism in Exercise 1. _____

6. Multiply your answer to Exercise 1 by $\frac{1}{2}$. Fill the measuring cup with this amount of rice. Make a rectangular prism that holds approximately the amount of rice in the measuring cup. On a separate sheet of paper, sketch a net for this prism. Label all dimensions of the prism, in centimeters. If you like, compare your net to the nets of other students.

7. Using the formula for area of a rectangle, find the area of each face of the prism you constructed in Exercise 6. What is the sum of the six areas? _____

8. If you write a fraction comparing the volume of the prisms in Exercises 6 and 1, the fraction is $\frac{1}{2}$. Write a fraction in simplest form showing how the sums of the areas of the faces compare (your answers for Exercises 7 and 2). Use the answer for Exercise 7 as the numerator of your fraction. How does this fraction compare to $\frac{1}{2}$? Explain whether you think the prism in Exercise 6 is "half as big" as the prism in Exercise 1. _____

Long-Term Project

The "Rice" Report, Chapter 6, page 3

Use a net to construct a triangular prism. Copy the net onto poster board or cardboard using the given dimensions, in centimeters.

After you draw the net, fold it along the solid lines. Tape the edges together to form a rectangular prism. Leave two top edges open so that you can lift the top.

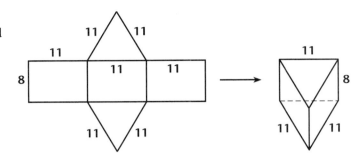

9. Pour rice into the prism until it is completely filled. (If you are using poster board, try to hold the sides straight so that you get an accurate measurement of volume.) Pour the rice into a measuring cup.

   Approximately how many mL of rice does your prism hold? _____

10. Using the formulas for area of a rectangle and area of a triangle that you learned in Chapter 5, find the area of each face of the prism. What is the sum of the five areas?

   _____

11. Multiply your answer to Exercise 9 by 2. Fill the measuring cup with this amount of rice. Make a triangular prism that holds approximately the amount of rice in the measuring cup. On a separate sheet of paper, sketch a net for this prism. Label all dimensions of the prism, in centimeters. If you like, compare your net to the nets of other students.

12. Using the formulas for area of a rectangle and area of a triangle, find the area of each face of the prism you constructed in Exercise 11. What is the

   sum of the five areas? _____

   _____

13. If you write a fraction comparing the volumes of the prisms in Exercises 11 and 9, the fraction is $\frac{2}{1}$. Write a fraction in simplest form showing how the sums of the areas of the faces compare (your answers for Exercises 12 and 10). Use the answer for Exercise 12 as the numerator of your fraction. How does this fraction compare to $\frac{2}{1}$? Explain whether you think the prism in Exercise 11 is "twice as big" as the prism in Exercise 9.

   _____

   _____

Geometry   Long-Term Project   23

NAME _____ CLASS _____ DATE _____

Long-Term Project
The "Rice" Report, Chapter 6, page 4

14. Multiply your answer to Exercise 9 by $\frac{1}{2}$. Fill the measuring cup with this amount of rice. Make a triangular prism that holds approximately the amount of rice in the measuring cup. On a separate sheet of paper, sketch a net for this prism. Label all dimensions of the prism in centimeters. If you like, compare your net to the nets of other students.

15. Using the formulas for area of a rectangle and area of a triangle, find the area of each face of the prism. What is the sum of the five areas? _____

16. If you write a fraction comparing the volumes of the prisms in Exercises 14 and 9, the fraction is $\frac{1}{2}$. Write a fraction in simplest form showing how the sums of the areas of the faces compare (your answers for Exercises 15 and 10). Use the answer for Exercise 15 as the numerator of your fraction. How does this fraction compare to $\frac{1}{2}$? Explain whether you think the prism in Exercise 9 is "half as big" as the prism in Exercise 14.

Part II: Make a display. To complete this project, follow these steps.

Step 1: Look at Exercises 1–8. For Exercise 1, use a net for a cube that is 8 centimeters on a side. Repeat Exercises 2–8 by substituting the word "cube" for "rectangular prism." Construct the cubes and draw nets to use in your display.

Step 2: Prepare an attractive display of your prisms. (A large cardboard box with the top and one side removed would make a good background for the prisms.) If you like, use color to make your display more appealing.

Step 3: Report any interesting results that you saw as you compared each prism's volume with the sum of the areas of its faces. Include what you think it means for one prism to be "twice as big" or "half as big" as another prism. (Your findings and sketches could be written on paper and attached to the inner sides of the box you use for the display.)

Step 4: Ask another student to critique your display.

NAME _____ CLASS _____ DATE _____

Long-Term Project

Swimming, Swimming in Our Swimming Pool, Chapter 7, page 1

Imagine this: it is summer, and you have your own swimming pool in your back yard — the cool, blue water inviting you to take a dip! People who want a swimming pool have several options in above-ground and in-ground models. Each has its advantages and disadvantages. In this project, you will look at the dimensions, surface area, and volume of differently shaped swimming pools. Then you will design your own swimming pool.

Part I: Consider the dimensions, surface area, and volume of swimming pools. Answer the questions, drawing diagrams as needed. This will help you plan your own swimming pool design.

Michael has a circular, above-ground pool in his back yard. His parents need to know the volume of the pool for several reasons. First, the city charges them by the gallon for water and they want to know how much water will be needed to fill it up. Second, to maintain the quality of the water, they need to use different chemicals in the pool. The amount of chemical depends upon the amount of water in the pool.

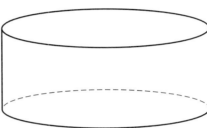

1. Michael's pool has a diameter of 16 feet. What is the surface area, in square inches, of the exposed surface of the water? Why might Michael be interested in the surface area of the exposed water?

   _____

2. Michael's pool has a height of 4 feet. However, the pool can only be filled to a depth of 42 inches. What is the volume of the pool in gallons? (One gallon is 231 cubic inches.)

   _____

3. Explain how you calculated the volume in gallons. _____

   _____

When Michael and his parents were shopping for pools, they saw an above-ground model as shown in the diagram (top view). The center portion was a rectangle and the ends were each semicircles. The rectangle portion was designed so that the length of the rectangle was twice the width of the rectangle.

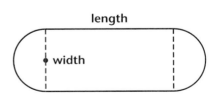

4. This pool can only be filled to a depth of 38 inches. There is a choice of the 8,000 gallon model or the 10,000 gallon model. What are the dimensions, in feet, of each model? Show your calculations.

   _____

   _____

Geometry   **Long-Term Project**

NAME _____ CLASS _____ DATE _____

Long-Term Project

Swimming, Swimming in Our Swimming Pool, Chapter 7, page 2

Michael's friend, Ben, wants a pool. His parents are looking at a new type of in-ground pool. For this pool, you dig the pool and use cement for the bottom and walls. Then, you use a durable vinyl liner. The liner comes in a variety of patterns that look like tile. The pool is dug so that it is the same depth everywhere. Ben and his family are interested in two models. The first one is an irregular shape shown on grid paper. The second one is shaped like a regular hexagon. Both pools can be filled to a depth of 4 feet.

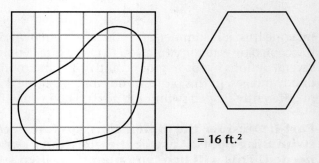

5. Find the approximate surface area, in square feet, of the exposed water in the irregular-shaped swimming pool. Explain your method for finding the

   area. Each square of the grid represents 16 square feet. _____

   _____

   _____

6. Find the approximate volume, in gallons, of the irregular-shaped pool. _____

   _____

7. Below, make a three-dimensional sketch of the hexagonal pool.

8. The hexagonal pool has a volume of 10,000 gallons. What is the length,

   in feet, of each side of the hexagonal pool? _____

   _____

9. Explain how you calculated the length of each side. _____

   _____

   _____

26 Long-Term Project   Geometry

Long-Term Project

Swimming, Swimming in Our Swimming Pool, Chapter 7, page 3

If you suddenly became very rich (maybe by winning the lottery!), you could buy a house with a huge yard and put in a pool that is as big as the pools found in city parks. Large pools usually vary in depth. Top and side views of the pool are shown. As you can see, the depth of the pool varies. The pool can be divided into three trapezoidal prisms as shown. (Drawing is not to scale.)

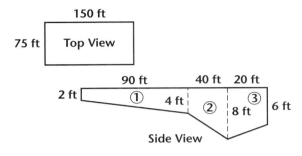

10. Below, sketch a diagram of each of the three trapezoidal prisms. Label the dimensions, in feet.

11. Find the surface area of the exposed water, in square feet. _____

12. Find the volume of the swimming pool, in gallons. _____

Michael and Ben's friend, Dane, practice for a diving team in this pool. This swimming pool has a special diving area. Top and side views are shown. The swimming area varies in depth as shown. The diving area can be thought of as a rectangular prism joined to an inverted (upside down) pyramid. (Drawing is not to scale.)

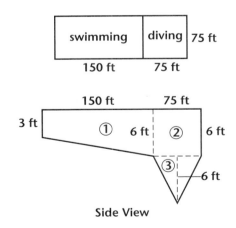

13. Below, sketch a diagram of the trapezoidal prism that represents the regular swimming area. Then sketch a diagram of the rectangular prism and inverted pyramid that represent the diving area. Label the dimensions of each, in feet.

NAME _____ CLASS _____ DATE _____

Long-Term Project
Swimming, Swimming in Our Swimming Pool, Chapter 7, page 4

14. Find the volume of the swimming pool, in gallons. _____

If you think the swimming pool in Exercise 14 was large, what do you think about this pool? According to The Guinness Book of Records, the largest swimming pool in the world is located in Casablanca, Morocco. It is filled with sea water. The dimensions are 1,574 feet by 246 feet.

15. Find the surface area of the water in acres. (Recall that 1 acre = 43,560 square feet.) Show your calculations.

_____

_____

16. Suppose the pool had a uniform depth of 4 feet. Find the volume of the swimming pool, in gallons.

_____

17. If this pool was circular and of uniform depth, what might be the dimensions? (You will need to select an appropriate depth for the water.)

_____

_____

Part II: Design your own swimming pool. To complete this project, follow these steps.

Step 1: Design a swimming pool. Provide a diagram of the top view of the pool and a diagram of the side view of the pool. Label the dimensions. The pool must be either an irregular shape or be a combination of two or more polygons.

Step 2: Find the surface area of the exposed water for your swimming pool. Show the calculations you used to find the surface area.

Step 3: Find the volume of your swimming pool. Show the calculations you used to find the volume.

Step 4: Visit a store that specializes in swimming pools. Research the various types available. If possible, obtain prices for the pools. Find out the advantages and disadvantages of above-ground and in-ground models.

Step 5: Make a brochure featuring your swimming pool. Include the diagrams of the pool and information about the pool, such as advantages and disadvantages.

Long-Term Project
Similar People, Chapter 8, page 1

You have learned that two polygons can be similar, that is, one polygon is the image of another by a dilation. Does similarity apply to humans? Are people similar? Are ratios comparing measurements of various "people parts" approximately the same? In this project, you will look at human ratios. You will also determine whether a doll, action figure, or figurine has the same ratios as humans.

Part I: Look at human ratios. Make measurements and construct tables as directed in the following exercises. These questions will help you to complete Part II of the project.

You will need the following materials:
   a cloth measuring tape with centimeter units
   several sheets of paper and some tape
   a doll or action figure that is similar to a human and can be easily measured
   centimeter grid paper

1. On a separate sheet of paper, prepare a table similar to the one below. You will need to have ten rows for entering the measurements you will make in Exercise 2.

Table 1

Name	Height	Foot Length	Wing Span	Navel	Tibia	Neck	Wrist

2. Use the following information to make measurements. Have someone measure you. Then, measure nine other people. Round all of your measurements to the nearest 0.5 centimeters. All measurements should be made without shoes. Record the measurements in the table you made in Exercise 1.

Height: Attach a piece of paper to the wall with tape at approximately the height of the person to be measured. Mark the person's height on the paper with tape or pencil. Use the measuring tape to measure from the floor to the mark.

Foot length: Stand on paper that is longer than the length of your foot. Make a mark at the back of your heel and at the tip of the longest toe. Use a measuring tape to measure this length.

Wing span: Have the person stretch their arms out to both sides. Measure across his/her back from the middle finger tip on the left hand to the middle finger tip on the right hand.

Navel to floor: Measure the length from the bottom of the foot to the navel.

Tibia: Measure the length from the ankle bone to the top of the tibia bone located on the outside of the knee cap.

Neck: Wrap the measuring tape loosely around the neck.

Wrist: Wrap the measuring tape around the widest part of the wrist.

Geometry — Long-Term Project

NAME _____ CLASS _____ DATE _____

Long-Term Project
Similar People, Chapter 8, page 2

3. On a separate sheet of paper, prepare a table similar to Table 2. You will need to have ten rows for entering ratios. Using the measurements in Table 1, write the ratios shown in Table 2 in fraction form. Then, change each fraction to a decimal, rounding to the nearest hundredth.

Table 2

Name	Wing Span:Height	Foot Length: Height	Navel:Height	Tibia:Height	Wrist:Neck

4. Look at your table of ratios. Compare the ratios in each column. Which ratios are approximately the same? _____

Calculate the average ratio for the following measurements of the ten people.

5. wing span:height _____

6. foot length:height _____

7. navel:height _____

8. tibia:height _____

9. wrist:neck _____

10. According to The Guinness Book of Records, the tallest man in medical history was Robert Wadlow, born in February of 1918. His height was recorded as 8 feet 11.1 inches. Using the average ratios you calculated in Exercises 5–9, estimate Robert's wing span, foot length, navel to floor length, and tibia length.

_____

_____

11. Explain how you used the average ratios to calculate Robert's measurements. _____

_____

12. Calculate measurements for Robert's wrist and neck. Explain how you calculated his measurements. _____

_____

NAME _____ CLASS _____ DATE _____

Long-Term Project
Similar People, Chapter 8, page 3

13. According to The Guinness Book of Records, the shortest woman was Pauline Musters, born in February of 1876. Her height was recorded as 24 inches. Using the average ratios you calculated in Exercises 5–9, estimate Pauline's wing span, foot length, navel to floor length, and tibia length.

_____

_____

14. Explain how you used the average ratios to calculate Pauline's measurements.

_____

_____

15. Calculate measurements for Pauline's wrist and neck. Explain how you calculated these

    measurements. _____

    _____

    _____

16. When you measured ten people, you used centimeters as the unit of measurement. Robert and Pauline's heights were given in feet and inches. How does the unit of measure affect the ratios of

    the body measurements? _____

    _____

Obtain a doll, action figure, or figurine that appears to be similar to a human. Make sure that it can be easily measured. In the following exercises, the item you choose will be referred to as a doll.

17. Using centimeters, measure and record only the height of your doll. Find your measurements in

    Table 1. What is the ratio of doll height: your height? _____

18. Using your measurements, estimate the doll's foot length, navel to floor length, tibia length,

    wrist, and neck measurements. _____

    _____

    _____

19. Explain how you calculated your doll's measurements. _____

    _____

    _____

Geometry   Long-Term Project

NAME _____ CLASS _____ DATE _____

Long-Term Project
Similar People, Chapter 8, page 4

20. Now measure the doll's foot length, navel to floor length, tibia length, wrist, and neck. Is the doll similar to you? Explain your reasoning. _____
_____
_____

21. Get a sheet of centimeter grid paper. Without shoes, trace your foot on the grid paper. Estimate the area of your foot by using the grid paper. Explain your method of estimation. _____
_____
_____

22. Estimate the area of your doll's foot by using the information in Exercise 21. Explain your method of estimation. _____
_____
_____

23. Get a sheet of centimeter grid paper. Without shoes, trace the doll's foot on the grid paper. What is the area of the doll's foot? How does it compare to your estimate from Exercise 22?
_____
_____

Part II: Compare a doll or action figure to a human. To complete this project, follow these steps.

Step 1: Measure five to ten young children or babies. Make a table of the measurements.

Step 2: Make a table of ratios for the children or babies.

Step 3: Compare the ratios for children or babies to the ratios you calculated in **Part I**.

Step 4: Obtain a doll, action figure, or figurine that represents a young child or baby. Determine whether the children or babies you measured are similar to this doll.

Step 5: Make a brochure or poster presenting your findings about human ratios. Include tables and any other information that will help explain your findings.

NAME _____ CLASS _____ DATE _____

Long-Term Project

Excursions into the History of Pi, Chapter 9, page 1

In Chapter 5, you investigated the approximate value of a special ratio, π. You determined that the value was approximately equal to 3.14. This special number, π, is not used exclusively in geometry. It appears in other areas of mathematics such as statistics, number theory, and actuarial theory. The number π has a long history. In this project, you will look at several developments in the history of π. You will also try several methods for estimating the value of π.

Part I: Look at the history of π. Complete the questions to help you understand the various methods used historically for estimating the value of π.

You will need the following materials:
  paper, compass, protractor, ruler
  geometry drawing software (optional)

Early Egyptian mathematicians wanted to find a square with the same area as a given circle. The ancient Rhind papyrus (c. 1650 B.C.) contains problems about this idea.

1. "Construct a square whose side length equals the diameter of the given circle. Find eight-ninths of the length. Square this number. The result is the area of the circle." To demonstrate this, draw a circle with diameter 10 centimeters on a separate paper. Construct a square on the diameter whose side length is the same as the diameter.

2. Multiply the value for the length of the side of the square in Exercise 1 by $\frac{8}{9}$. To find the area of the circle, square this value. What is the area of the circle using this calculation? _____

   _____

3. Use the formula you learned in Chapter 5 to find the area of the circle you drew in Exercise 1. How does this value compare to your answer to Exercise 2? _____

   _____

4. Draw two more circles of different sizes on your paper and construct squares on the diameters. Find the area of the circles using both the formula from the Rhind papyrus and the formula from Chapter 5. Describe your results. _____

   _____

In 1865, a Polish Jesuit, Adamas Kochansky, designed a fairly simple construction with straightedge and compass for approximating the value of π. Complete the following exercises to model his procedure.

5. On a separate sheet of paper, follow these steps and refer to the diagrams as needed.

   a. Use a compass to draw a circle of radius one inch. (The circle at right shows a radius of one-half inch.) Draw a vertical diameter for the circle. Then construct a tangent at each end of the diameter as shown in the diagram.

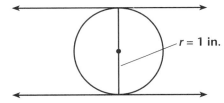

Geometry   Long-Term Project **33**

Long-Term Project

Excursions into the History of Pi, Chapter 9, page 2

b. Construct a 30° angle as shown. Label the intersection of the side of the angle and the tangent as point A.

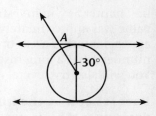

c. Label the intersection of the diameter and the other tangent as point B. Extend the tangent and mark point C such that $\overline{BC}$ is the length of three radii of the circle.

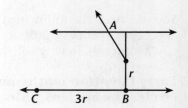

d. Draw $\overline{AC}$. Measure, to the nearest sixteenth of an inch, the length of $\overline{AC}$.

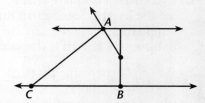

6. How does the length of $\overline{AC}$ in Exercise 5 compare to the value of π? _____

7. Repeat Exercise 5 using circles of radius two inches and three inches. Describe your results.

In 1674, a German mathematician, von Leibniz, found a formula for π that was not dependent upon geometric constructions. His formula was $\pi = 4 \cdot \left(1 - \frac{1}{3} + \frac{1}{5} - \frac{1}{7} + \frac{1}{9} - \frac{1}{11} + \ldots\right)$, where the three dots mean to continue the pattern.

8. Use a calculator to find a value for π using just the portion of the formula shown. (Don't continue the pattern.) What is your estimate for π using this formula?

9. Extend the pattern until the last term is $\frac{1}{23}$. Use a calculator to find this estimate for π.

NAME _____ CLASS _____ DATE _____

Long-Term Project
Excursions into the History of Pi, Chapter 9, page 3

Archimedes (c. 287–212 B.C.), a Greek mathematician, first performed what is now called the "classical method" for calculating π. In 1579, Viete, a French mathematician, used this method to obtain a better estimate for π. In 1610, van Ceulen, a German mathematician, obtained yet a better estimate for π. Complete the following exercises to model the work of early mathematicians.

10. On a separate sheet of paper, make a table similar to the one below or use spreadsheet software.

Name of Polygon	Radius of Circle	Perimeter of Inscribed Polygon	(Perimeter of Inscribed Polygon) ÷ (2 · radius)	Perimeter of Circumscribed Polygon	(Perimeter of Circumscribed Polygon) ÷ (2 · radius)
triangle					
square					
pentagon					
hexagon					
heptagon					
octagon					
nonagon					
decagon					
11-gon					
dodecagon					

11. On a separate paper, follow these steps and refer to the diagrams as needed. You will construct an equilateral triangle inscribed in a circle, meaning it has all vertices on the circle.

  a. Use a compass to draw a fairly large circle. Label the center O. Draw in one radius, $\overline{OA}$. Measure $\overline{OA}$ and enter this value in column 2 of the table you made in Exercise 10.

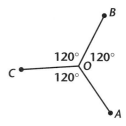

  b. Since the triangle has 3 sides, divide 360° by 3. The result is 120°. Using $\overline{OA}$ as one side of the angle, construct a 120° central angle. Then construct two more 120° central angles. Label the intersection points of the sides of the angles with the circle as shown.

  c. Draw segments $\overline{AB}$, $\overline{BC}$, and $\overline{AC}$ to form a triangle. △ABC is inscribed in circle O, since all vertices are on the circle.

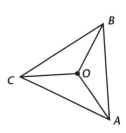

  d. Find the perimeter of △ABC. Record the value for the perimeter in column 3 of the table you made in Exercise 10.

  e. Divide the value in column 3 by 2 times the value in column 2. Record this value in column 4 of the table you made in Exercise 10.

Geometry   Long-Term Project

NAME _____ CLASS _____ DATE _____ Geometry

Long-Term Project
Excursions into the History of Pi, Chapter 9, page 4

12. Follow these steps and refer to the diagram as needed. You will construct an equilateral triangle circumscribed about the circle you constructed in Exercise 11. Each side of a circumscribed polygon is tangent to the circle.

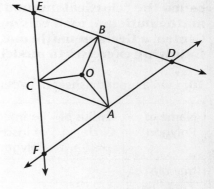

  a. At each point A, B, and C construct a tangent to circle O. Label the points of intersection of the tangents as shown. Connect the points to form a triangle. △DEF is circumscribed about circle O, since each side of the triangle is tangent to the circle.

  b. Find the perimeter of △DEF. Record the value for the perimeter in column 5 of the table you made in Exercise 10.

  c. Divide the value in column 5 by 2 times the value in column 2. Record this value in column 6 of the table you made in Exercise 10.

13. Explain how you constructed the tangents to circle O in Exercise 12, Step **a**.

_____

14. Use theorems from Lessons 9.3 and 9.4 to explain how you can be sure

that all the angles of △ABC and △DEF are 60° angles. _____

_____

_____

Part II: Model methods for calculating π. To complete this project, follow these steps.

  Step 1: For each polygon listed in the table you made in Exercise 10, inscribe that polygon in a circle. Then, circumscribe the same type of polygon around the same circle. Follow the steps listed in Exercises 11 and 12. Remember that you will divide 360° by the number of sides in the polygon to find the central angle. All polygons should be regular.

  Step 2: Record the radius of each circle and the perimeters of the polygons in the table from Exercise 10. Also, fill in columns 4 and 6. Observe what happens to the values in columns 4 and 6 as the number of sides of the polygon increases.

  Step 3: Prepare a booklet or poster showing the various methods you used to calculate π in this project. Include diagrams and tables as needed.

36   Long-Term Project   Geometry

NAME _____ CLASS _____ DATE _____

Long-Term Project
Congratulations! You're Officially a Mathematician, Chapter 10, page 1

In Chapter 5, you learned a general formula for finding the area of any regular polygon, $A = \frac{1}{2}ap$, where a is the length of the apothem and p is the perimeter. However, when you draw a regular polygon, you usually know the length of the side but not the length of the apothem. Are there formulas for specific regular polygons in which you only need to know the length of the side? How do mathematicians find area formulas? In this project, you will find formulas for the area of specific regular polygons. You will get the opportunity to experience what mathematicians do when they need to find a new formula. Specific formulas are very handy for using in spreadsheets where you need to find the area of quite a few polygons with various side lengths.

Part I: Find formulas. Complete the following exercises to help you learn how to write your own formulas for the areas of specific regular polygons.

You have seen specific formulas for the areas of triangles and squares. So let's start by finding a formula for the area of a regular pentagon. When you have finished these exercises, you will just need to substitute the length of half the side of any regular pentagon into the formula to calculate the area.

First, carefully draw a regular pentagon on a separate sheet of paper. Have each side of the pentagon measure 4 inches. Bisect the two base angles of the pentagon as shown. Where the bisectors intersect, construct a perpendicular to the base. Label the pentagon as shown. The apothem is $\overline{OB}$.

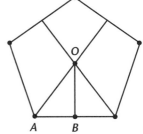

1. On your pentagon, label the degree measures of $\triangle AOB$. What are the measures of the three angles?

_____

2. By construction, what is the length of $\overline{AB}$? _____

3. Express the length of $\overline{OB}$ in terms of the tangent of $\angle OAB$. (Be sure to substitute the actual degree measure of $\angle OAB$ into the expression.) _____

4. What is the perimeter of the pentagon, in inches? _____

5. Substitute your answers for Exercises 3 and 4 into the formula $A = \frac{1}{2}ap$. Simplify the result to get the area of the pentagon, in square inches. Round your answer to the nearest hundredth. _____

_____

_____

Geometry   Long-Term Project

Long-Term Project
Congratulations! You're Officially a Mathematician, Chapter 10, page 2

6. Now you need to generalize the formula. Let x be the length of $\overline{AB}$ in the pentagon you drew. (Then each side of the pentagon will be length $2x$.) Write an expression for the perimeter, p, using x. Write an expression for the length of $\overline{OB}$ or the apothem, a, using x. (See your answer to Exercise 3.) Now substitute these expressions into the formula $A = \frac{1}{2}ap$. Simplify the result. Congratulations — you now have a generalized formula for the area of any regular pentagon,

given the length of half the side. The formula is: _____

Would it be possible to write a different formula for the area of a pentagon? The next formula is going to be more difficult to find. It is based upon The Law of Cosines from Chapter 10 and a formula called Heron's Formula discovered by Heron, an ancient Greek mathematician. Heron's formula is: the area of any triangle $= \sqrt{s(s-a)(s-b)(s-c)}$, where s is half the perimeter of the triangle and a, b, and c are the lengths of the three sides.

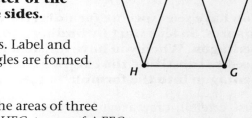

Draw another regular pentagon with side length of 4 inches. Label and connect the vertices as shown. You can see that three triangles are formed. They are $\triangle DEH$, $\triangle HEG$, and $\triangle FEG$.

7. The area of pentagon $DEFGH$ can be found by adding the areas of three triangles: area of pentagon = area of $\triangle DEH$ + area of $\triangle HEG$ + area of $\triangle FEG$.

   On a separate sheet, draw each triangle individually, as shown. On the diagram, label the degree measures of all three triangles. Don't rely on measurement; use geometric properties to find the measures.

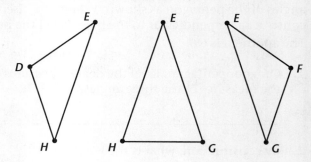

8. Which two triangles are congruent? How do you know?

   _____

9. Look at triangle $\triangle DEH$. To use Heron's formula for the area of this triangle, you must find the length of $\overline{HE}$. Let's generalize the formula now. Let y be the length of $\overline{DE}$ which will then also be the length of $\overline{DH}$. Use the Law of Cosines to write an expression for the length of $\overline{HE}$. Simplify the expression. (Recall that $\sqrt{y^2} = y$.) Round decimals to the thousandths place.

   _____

38 Long-Term Project — Geometry

NAME _____ CLASS _____ DATE _____

Long-Term Project

Congratulations! You're Officially a Mathematician, Chapter 10, page 3

10. Now you have the lengths of the three sides (called a, b, and c by Heron) of $\triangle DEH$. They are y, y, and your answer for Exercise 9. Label the sides of $\triangle DEH$ with the side lengths. Find the half perimeter, s, of the triangle. $s =$ _____

11. To find the area of $\triangle DEH$. substitute the expressions for the lengths of s, a, b, and c into Heron's formula. Simplify the result rounding decimals to the thousandths place. Area =

_____

Your answer for Exercise 11 is a generalized formula for the area of $\triangle DEH$ and the area of $\triangle FEG$, since they are congruent triangles. Now all you need to find is the area of $\triangle HEG$.

12. Recall that the length of $\overline{HG}$ is the same as the length of $\overline{DE}$ and $\overline{DH}$ which is also y. In Exercise 9 you wrote a generalized expression for the length of $\overline{HE}$ which is also a side of $\triangle HEG$. Notice that the lengths of $\overline{HE}$ and $\overline{EG}$ are the same. Label the $\triangle HEG$ with the expressions for the side lengths. Find the half perimeter, s, of the triangle. $s =$ _____

13. To find the area of $\triangle HEG$, substitute the expressions for the lengths of s, a, b, and c into Heron's formula. Simplify the result rounding decimals to the thousandths place. Area = _____

_____

14. Find your answers to Exercises 11 and 13. Substitute these expressions for the areas of $\triangle DEH$, $\triangle HEG$, and $\triangle FEG$ into this statement: area of pentagon = area of $\triangle DEH$ + area of $\triangle HEG$ + area of $\triangle FEG$. Simplify the result, rounding decimals to the thousandths place. Area of pentagon =

_____

15. To see if your formula really works, substitute 4 (the length of the side of the pentagon you drew at the beginning of this project) for y into your formula from Exercise 14. How does the answer compare to the area of the pentagon you calculated in Exercise 5? _____

_____

16. As a final check of your formulas, refer to your drawing of a pentagon with side length 4 inches. Estimate the length of the apothem and find the perimeter of the pentagon. Substitute these values into the formula $A = \frac{1}{2}ap$ and find the area in square inches. How does the result compare to your answers for Exercises 5 and 15? _____

_____

Geometry   Long-Term Project   39

Long-Term Project

Congratulations! You're Officially a Mathematician, Chapter 10, page 4

If you completed the project for Chapter 9, you will recall that you needed to find the perimeter of a regular polygon inscribed in a given circle. You know the radius of the circle, but not the length of each side of the polygon. Trigonometric ratios can help you write a formula for the perimeter if all you know is the radius of the circle.

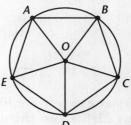

On a separate sheet of paper, construct a circle whose radius is 4 inches. Inscribe a regular pentagon in the circle. (If you have forgotten how to inscribe a pentagon, refer to the Chapter 9 Project.) Label your construction as shown. In your construction the length any segment from the center to a vertex will be equal to the radius, or 4 inches.

17. To find the perimeter of the pentagon, you need to know the length of one side, for example, the length of AE. To find this length, you need to know the angle measures of the five congruent triangles. Choose $\triangle AEO$. Make a sketch of $\triangle AEO$. Find the angle measures of $\triangle AEO$ and label them on your diagram.

18. On the diagram of $\triangle AEO$ you made in Exercise 17, label the midpoint of $\overline{AE}$ as point F. Draw $\overline{OF}$. Find the angle measures of $\triangle AOF$ and label them on your diagram.

19. Use the sine or cosine ratio to find the length of $\overline{AF}$. What is the length of $\overline{AF}$? _____

20. What is the perimeter of the pentagon inscribed in the circle of radius 4 inches? Round your answer to the nearest tenth. _____

21. Now you need to write a generalized formula for the perimeter of a regular pentagon inscribed in the circle of radius z. Use your answers to Exercises 19 and 20 to write the formula: Perimeter = _____

22. As a final check of your formula, refer to your drawing of a pentagon inscribed in the circle of radius 4 inches. Measure the length of the sides and find the perimeter of the pentagon. How does the result compare to your answer to Exercise 20? _____

Part II: Find and write your own formulas. To complete this project, follow these steps.

Step 1: Select two of the following polygons—hexagon, octagon, nonagon, decagon, or dodecagon—and draw a diagram of each polygon. The polygons must be regular.

Step 2: Use these diagrams and what you learned in **Part I** to write <u>two</u> different formulas for each polygon's area, given either the length of one-half the side or the whole side. Write one formula for each polygon's perimeter given the radius of the circle in which it is inscribed.

Step 3: Prepare a poster or brochure presenting your formulas. Report any patterns you found.

Long-Term Project
A Fractal Cut-Up, Chapter 11, page 1

In Lesson 11.6, you learned about fractals, geometric objects that exhibit some type of self-similarity. In this project, you are going to make your own three-dimensional fractal by measuring and cutting. You will also investigate some interesting characteristics of this fractal. For this project, you will need at least 5 sheets of printer paper (8.5 by 11 inches), sheets of paper for making tables, an inch ruler, scissors, and colored pencils. Spreadsheet software is optional for the tables.

Part I: Construct your own fractal. Follow the steps given and answer the questions. Save all of your paper constructions; they will help you complete Part II of this project.

1. Make a table using the headings shown. Call it Table 1. Leave room for at least 10 rows.

Stage Number	Placement of cut: one-fourth length of folded rectangle	Length of cut: one-half width of folded rectangle	Dimensions of inner rectangle	Inner rectangle: ratio of length to width

2. Cut four sheets of paper so that each sheet measures 8 inches by 10 inches. Label them Stage 1 through Stage 4. Each sheet will represent a stage, or iteration, of the fractal. Select Stage 1 and follow these steps. Refer to the diagrams as needed. Record a 1 in column 1 of Table 1.

   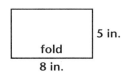

   a. Fold the sheet of paper so that it is a rectangle with length 8 inches and width 5 inches. Place the fold as shown in the diagram.

   b. Find one-fourth of the length of the rectangle, or 0.25 · 8 = 2 inches. Write 2 in column 2 of Table 1. Find half the width of the rectangle, or 0.5 · 5 = 2.5 inches. Write 2.5 in column 3 of Table 1. Draw two line segments as shown such that each segment is 2 inches from an end of the rectangle and is 2.5 inches in length from the fold. Carefully and accurately cut through both layers of paper on the line segments.

   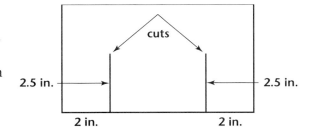

   c. Fold the cut section up to the top edge of the paper and crease sharply with your fingernail or other object. You now have an inner rectangle (shaded). Its dimensions should be 4 inches by 2.5 inches. Record this in column 4 of Table 1. In column 5, find the ratio, in decimal form, of length to width for the inner rectangle.

   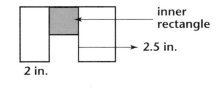

3. What is the ratio you found? Is it close to any special ratio you have studied in Chapter 11?

4. Unfold the sheet of paper. Refold all creases sharply so that they will bend in either direction. The two sections of the 8 inch by 10 inch paper should be positioned at a 90 degree angle, as shown, so that a box-like structure with open ends "pops up." Set Stage 1 aside for later.

   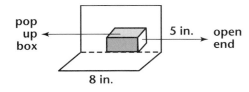

Geometry   **Long-Term Project**

Long-Term Project
A Fractal Cut-Up, Chapter 11, page 2

5. Select the Stage 2 sheet. Record a 2 in row 2 column 1 of Table 1. Make another Stage 1 fractal by repeating the steps in Exercise 2, but do not unfold the sheet as described in Exercise 4.

   a. Look at the inner rectangle which measures 4 inches by 2.5 inches. Find one-fourth of the length and one-half of the width and record in Table 1. Measure and cut two segments similarly to the way you did in Exercise 2. Fold up the next inner rectangle as you did in Exercise 2c. Fill in columns 4 and 5 in Table 1.

   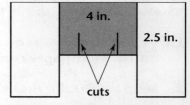

   b. Unfold the sheet of paper and refold the creases as in Exercise 4. The Stage 2 fractal should have more than one pop-up box.

6. Describe the Stage 2 fractal. Include the number and sizes of the pop-up boxes. _____

7. Select the Stage 3 sheet. Make another Stage 2 fractal by repeating the steps in Exercise 2, but do not unfold the sheet. Then repeat 5a, but not 5b. On the next inner rectangle, draw segments using the same process as you did earlier — the segments are one-fourth of the way from each end and are one-half the width of the inner rectangle. Fill in the row for Stage 3 in Table 1. Unfold the third sheet of paper and refold the creases. Describe the Stage 3 fractal, including the number and sizes of the pop-up boxes. _____

8. Select the Stage 4 sheet. Use this sheet to make a Stage 4 fractal. Fill in Table 1 for Stage 4. Describe the Stage 4 fractal, including the number and sizes of the pop-up boxes.

9. Imagine that you were able to continue with more sheets of paper. Look for patterns in Table 1. Without cutting any more fractals, fill in Table 1 through Stage 10. Describe any patterns.

10. Make a table using the headings shown. Call it Table 2. Leave room for at least 10 rows. Look at your four sheets of paper representing Stages 1 through 4 of the fractal. The Stage 1 fractal had 1 pop-up box. More pop-up boxes were added at each stage. Fill in this table with the number of boxes added at each stage and the total number of boxes at each stage. Continue the pattern to include Stage 10 of the fractal.

Table 2

Stage Number	Number of Boxes Added	Total Number of Boxes
1	1	1
2	?	1 + ? =

NAME _____ CLASS _____ DATE _____

Long-Term Project
A Fractal Cut-Up, Chapter 11, page 3

11. Describe the pattern you used to find the total number of boxes at any stage. _____

An interesting relationship can be seen when you investigate the surface area of the pop-up boxes of the fractal at each stage. Complete the following exercises to find the pattern and make a table of surface area.

12. Make a table using the headings shown. Call it Table 3. Leave room for at least 10 rows.

Table 3

Stage Number	New Surface Area Added	Total Surface Area	Total Surface Area at This Stage ÷ Surface Area at Stage 1

13. Surface area at any stage will be defined as any face of a pop-up box that is covered by paper. For example, look at Stage 1. You can see one pop-up box. The bottom, back, and the ends of the box formed are open. These 4 faces will not be considered in the surface area. The pop-up box has two rectangular faces that are covered by the paper. Another way to see this surface area is to flatten Stage 1. Color red the rectangle whose sides are the cuts and the creases. This will be the surface area of the Stage 1 fractal. What is the surface area of the Stage 1 fractal? _____

14. Fill in the first row of Table 3 with Stage 1, the surface area that was added (your answer to Exercise 13), the total surface area (same as column 2), and the total surface area at this stage ÷ surface area at Stage 1 (column 3 divided by 20). What is the value in column 4? _____

15. Get Stage 2. Fold the sheet flat. Color the exact same area red as you did in Exercise 13. Color any additional surface area blue. Find the area of the blue region. _____

16. Pop the boxes back out for Stage 2. Look at the two small pop-up boxes. What is the area of the 4 faces of the two small pop-up boxes? How does this compare to the area of the blue region you calculated in Exercise 13? _____

17. What is the amount of surface area that is added at Stage 2? Is it the answer from Exercise 15 or the answer from Exercise 16? _____

18. Fill in Table 3 for Stage 2. What is the value in column 4? _____

19. Get Stage 3. Fold the sheet flat. Color the exact same area red as you did in Exercise 13. Color the exact same area blue as you did in Exercise 15. Color any additional surface area at this stage yellow. What is the amount of surface area that is added at Stage 3? _____

Long-Term Project
A Fractal Cut-Up, Chapter 11, page 4

20. Fill in Table 3 for Stage 3. What is the value in column 4? _____

21. Get Stage 4. Repeat the coloring process to find the additional surface area. Color the added area green. Fill in Table 3 for Stage 4. What is the value in column 4? _____

22. Look for any patterns in Table 3. Fill in the table for Stages 5 through 10. Describe the patterns you found. _____

23. Now that you have constructed a fractal consisting of "pop-up" boxes, try this pattern to construct a different fractal. Use one sheet of paper and continue the process shown until the paper becomes too thick to cut and fold.

    a. Fold a sheet of paper in half with the fold at the bottom.

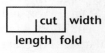

    b. Make a cut in the paper that is half the length of the rectangle and continues from the fold for half the width of the rectangle.

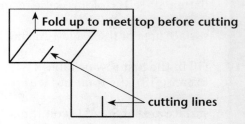

    c. Fold up the left side of the paper. Repeat the cutting process using half the length and width of the folded up rectangle. Cut both the left side and the right side on the lines that are shown.

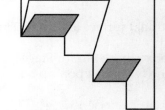

    d. Continue folding and cutting as long as possible. Unfold and crease sharply.

    e. Describe the fractal.

Part II: Make a display. To complete this project, follow these steps.

    Step 1: Collect your fractals from **Part I**. If you prefer, you can make new fractals by beginning with larger sheets of paper.

    Step 2: Collect your tables from **Part I**. If you made new fractals, make new tables as needed.

    Step 3: Make an attractive display of your fractals, tables, and findings. Report any interesting patterns in the fractals or the tables. If you used a spreadsheet, include the formulas you used to find the entries for Stages 5 through 10 in Tables 1, 2, and 3.

Long-Term Project
"Venn" Ever, Chapter 12, page 1

A useful tool for solving logic problems is a Venn diagram. John Venn (1834–1923) invented these diagrams to represent expressions in Boolean algebra. At about the same time, Leonard Euler (1707–1783) invented similar diagrams. Boolean algebra deals with sets and logic. In this project you will learn more about Boolean algebra, Venn and Euler diagrams, and sets. This will help you to determine the validity of logical arguments and write your own logical arguments. For this project, you will need 4 large rubber bands (approximately 16 inches in circumference), 4 smaller rubber bands, (approximately 10 inches in circumference) and 4 sheets of colored paper, one each of red, blue, yellow, and green.

Part I: Use Venn and Euler diagrams to represent sets, logical arguments, conjunctions, and disjunctions.

Euler diagrams can help you represent sets and subsets. Complete the following exercises.

1. From each sheet of colored paper, cut two squares — one about 1 inch on a side and the other about 2 inches on a side. From each sheet of paper, cut two equilateral triangles — one about 1 inch on a side and the other about 2 inches on a side. From each sheet of paper, cut two circles — one about 1 inch in diameter and the other about 2 inches in diameter. You should have 24 geometric figures. One characteristic of each figure is its shape such as square, circular, or

   triangular. What are other characteristics exhibited by the figures? _____

2. A *set* is a collection or group of objects. The objects in a set are called *elements*. An Euler diagram uses circles to represent sets. Using your figures from Exercise 1, one set is "the set of all squares." Sets are usually identified by letters. Another way to identify the set of squares is $A = $ {squares}. A description of the elements of the set or a listing of the elements is placed in brackets. You could also write $A = $ {small red square, large red square, small yellow square, large yellow square, small blue square, large blue square, small green square, large green square}. You can represent the set of squares using a large rubber band and the squares you made in Exercise 1. Form a circle with a rubber band and place all the squares inside.

   An Euler diagram representing set A would look like this:

   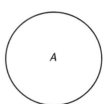

3. Get the two yellow squares and one smaller rubber band. Place the two yellow squares inside the smaller rubber band. You can let $B = $ {yellow squares}. Place set B inside the rubber band for set A. Notice that all elements of set B are contained in set A. A subset is a set such that all of its elements are contained in another set. The notation is $B \subset A$, read "B is a subset of A."

   A subset can contain fewer elements than the set or the same number of elements as the set. An Euler diagram representing a subset would look like the figure at right.

   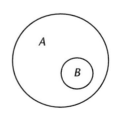

NAME _____ CLASS _____ DATE _____

Long-Term Project
"Venn" Ever, Chapter 12, page 2

4. Use your geometric figures and rubber bands to show 4 different cases of sets and subsets. On a separate sheet of paper, draw Euler diagrams for your cases. Describe each set and subset.

5. Sets can contain any type of object. They can also have a definite number of items, such as 8 squares, or the number of items can be infinite. For example, set C is specified as the set of all counting numbers or $C = \{1, 2, 3, ...\}$, where the three dots indicate that you just keep counting. This is an example of an infinite set. Write two examples of infinite sets.

_____

Euler diagrams, sets, and subsets can help you determine the validity of an argument. Complete the following exercises to appreciate the value of using Euler diagrams.

6. You can draw an Euler diagram to represent a statement with a premise and a conclusion. For example, consider the statement: "Every person who is 13 years old is a teenager." It is easiest to use this statement if it is rewritten as: "If a person is 13 years old, then he/she is a teenager."

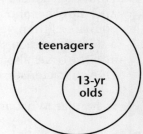

An Euler diagram for this situation is shown. A subset of the set of teenagers is the set of 13-year-olds.

The set of 13-year-olds is not the only subset of the set of teenagers. On a separate sheet draw an Euler diagram showing at least three other subsets of the set of teenagers.

7. Suppose you must determine the validity of a conclusion. Consider this modus ponens argument.

If a person is 13 years old, then she/he is a teenager. ⬅— Hypothesis
Michael is 13 years old.
Therefore, Michael is a teenager. ⬅— Conclusion
An Euler diagram would look like this, where the x represents Michael as a 13-year-old.

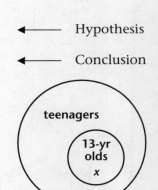

Since Michael is inside the set of teenagers, you can conclude that Michael is a teenager. Suppose that you are given that Michael is a teenager. Can you conclude that he is a 13-year-old? Explain.

_____
_____

NAME _____ CLASS _____ DATE _____

Long-Term Project
"Venn" Ever, Chapter 12, page 3

8. Determine whether each of the following is a valid conclusion. On a separate sheet of paper, draw an Euler diagram for each argument and explain why the conclusion is valid or invalid. If necessary, first write each statement as an if-then statement.

   a. If a living organism is a marsupial, then it is a mammal. A Tasmanian devil is a marsupial. ←—— Hypothesis

   Therefore, a Tasmanian devil is a mammal. ←—— Conclusion

   b. If a person wants to be assured a seat on a domestic airline flight, then he/she must check in at least one hour in advance. Ian checks in 24 hours in advance. ←—— Hypothesis

   Therefore, Ian can be assured a seat on a domestic airline flight. ←—— Conclusion

   c. Every bird lays eggs to reproduce. Canadian geese are birds. ←—— Hypothesis

   Therefore, Canadian geese lay eggs to reproduce. ←—— Conclusion

   d. In the United States, all states have a governor. Idaho has a governor. ←—— Hypothesis

   Therefore, Idaho is a state in the United States. ←—— Conclusion

   e. All children in the dance recital are at least 3 years old. Aislyn is in the dance recital. ←—— Hypothesis

   Aislyn is at least 3 years old. ←—— Conclusion

9. Look in newspapers, magazines, or books to find facts that can be written as if-then statements. On a separate sheet, write four of your own exercises similar to parts **a–e** in Exercise 8 using these facts. Then draw Euler diagrams and explain why the conclusions are valid or invalid. Include at least one invalid conclusion in the four exercises.

Venn diagrams can help you to represent conjunctions and disjunctions. Complete the following exercises to help you understand this use of Venn diagrams.

10. Select all the circles and all the red geometric figures you made in Exercise 1. Place all circles inside one large rubber band. This will be set *A*. Place all the red geometric figures inside a second large rubber band. This will be set *B*. Where did you put the red circles? Explain your reasoning.

_____

_____

_____

_____

Geometry   Long-Term Project

NAME _____ CLASS _____ DATE _____

Long-Term Project
"Venn" Ever, Chapter 12, page 4

11. The intersection of two sets A and B, written $A \cap B$, is the set of all elements that are common to both sets. For example, in Exercise 10, the set of red circles is $A \cap B$. A conjunction can be used to describe the intersection of the sets A and B. The conjunction for this situation would be: The geometric figure is red and the geometric figure is a circle.

   A Venn diagram would look like this. Overlap your rubber bands from Exercise 10 and place the circles and red figures in the correct positions. Usually, you shade the area representing the intersection of the sets. Describe the elements in $A \cap B$.

   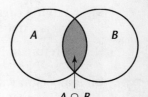

   _____

   _____

12. On a separate sheet, write four conjunctions using the geometric figures you made in Exercise 1. Draw a Venn diagram for each conjunction and identify the elements in the intersection of the sets. Use Exercises 10 and 11 as a guideline.

13. Select all the triangles and all the green geometric figures you made in Exercise 1. Overlap two large rubber bands. Place the green triangles inside the area represented by $A \cap B$. Place the rest of the triangles inside just set A. Place all the green figures inside just set B. How many figures are in set A, set B, or in both set A and set B?

   _____

The union of two sets A and B, written $A \cup B$, is the set of all elements that are in set A, set B, or in both set A and set B. For example, in Exercise 13, $A \cup B$ contains all triangles, all green figures, and all green triangles. A disjunction can be used to describe the union of the sets A and B. The disjunction for this situation would be: The geometric figure is a triangle or it is green.

A Venn diagram would look like this. The shaded region shows $A \cup B$ which includes all of set A, all of set B, and $A \cap B$.

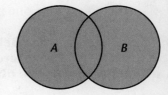

14. On a separate sheet, write four disjunctions using the geometric figures you made in Exercise 1. Draw a Venn diagram for each disjunction and identify the elements in the union of the sets.

Part II: Make a poster. To complete this project, follow these steps.

   Step 1: Collect your examples of using Euler diagrams for sets and logic from Exercises 4 and 9.

   Step 2: Collect your examples of using Venn diagrams for conjunctions and disjunctions from Exercises 12 and 14.

   Step 3: Make a poster using your diagrams from steps 1 and 2.

ANSWERS

Long-Term Project — Chapter 1

1. △ABC has no congruent sides or congruent angles. One angle is close to 90°.

2. intersection of Broadway and Grant

3. No, D is not the midpoint of $\overline{AC}$.

4. No, the angle formed by $\overline{BD}$ and $\overline{AC}$ ≠ 90°.

5. A′ is near Saints Peter and Paul Church, B′ is on Nob Hill, and C′ is between Jackson and Washington about one-half block from Powell.

6. A″ is between O'Farrell and Ellis about on Mason, B″ is between Clay and Sacramento near Taylor, and C″ is about on California and Joice.

7. Yes, a 180° rotation with center at approximately the intersection of Clay and Stockton

8. Answers may vary. A sample is given. Use a compass. Set the radius as DD″. Mark this distance using D as a center. Mark this distance using D″ as a center. Where they intersect near Pacific and Jones is the third vertex.

9. on Washington near the Cable Car Barn

10. Cable Car Barn

11. 020

12. 120

13. about 0.4 miles

14. about 0.1 miles

15. Pacific Union Club

16. Grace Cathedral

17. The angle measures remain the same. The actual distances in miles remain the same because you use a different scale depending upon the size of the map.

18. Answers may vary. You can write an solve the proportion $\frac{inches}{miles} = \frac{inches}{miles}$.

19. Answers may vary. Draw a line parallel to Hyde Street through the first location. Draw a line from the first location to the second location. Measure the angle between the two locations. Subtract 10 since Hyde Street is at 350, not 360.

Part II: Check students' work.

Long-Term Project — Chapter 2

1. If you are 15 years old, then you can apply for a regular license in Montana.

2. If you can apply for a regular license in Montana, then you are 15 years old.

3. No, you can be older and still apply.

4. Answers may vary. If you are 14 years old in South Dakota, then you can apply for a regular license. Converse: if you can apply for a regular license, then you are 14 years old in South Dakota. The converse is false because you can be older and still apply and you might live in a different state.

5. A state is in Group 1 if and only if the age for a regular license is 16 and the age for a learner's permit is 14.

6. A state is in Group 2 if and only if the age for a regular license and the age for a learner's permit is the same.

7. Answers may vary. A state is in Group 3 if and only if the age for a learner's permit is 15 or 15 plus some months and the age for a regular license is 16 or 16 plus some months. A state is in Group 4 if and only if the age for a learner's permit is 14, 15, or 16 and the age for a regular license is 17. A state is in Group 5 if and only if the age for a learner's permit is 15 or 16 and the age for a regular license is 18. A state is in Group 6 if and only if it is not in any other group.

ANSWERS

Group 3	Group 4	Group 5	Group 6
AL	CA	FL	DE
AZ	IL	GA	MT
CO	IA	IN	
CT	LA	MA	
KY	MD	NH	
ME	MI	NY	
MS	MN	PA	
MO	NE		
NV	NJ		
NC	OH		
OK	RI		
OR			
SC			
TN			
TX			
VT			
VA			
WA			
WV			
WI			
WY			

11.

STATEMENTS	REASONS
$m\angle 1 = m\angle 3$	Given
$m\angle 4 = m\angle 5$	Given
$m\angle 2 + m\angle 4 + m\angle 5 = 180°$	Given
$m\angle 1 + m\angle 2 + m\angle 3 = 180°$	Linear Pair Property
$m\angle 1 + m\angle 2 + m\angle 3 = m\angle 2 + m\angle 4 + m\angle 5$	Substitution Property
$m\angle 1 + m\angle 3 = m\angle 4 + m\angle 5$	Subtraction Property
$m\angle 1 + m\angle 1 = m\angle 4 + m\angle 4$	Substitution Property
$2m\angle 1 = 2m\angle 4$	Substitution Property
$m\angle 1 = m\angle 4$	Division Property

12.
```
      3        5           7
House  B. John's  Mall     Court House
```

13. 35°

14. 70°

15. 30°

16. 45°

17. 30°

18. 70°

19. 45°

20. Answers may vary. (1.) enrolled in a driver education class, (2.) met the age requirement, (3.) took the written test for a learner's permit, (4.) passed the test, (5.) practiced driving with a licensed driver and learned parallel parking, (6.) met the age requirement, (7.) took the driving test for a regular license, (8.) passed the driving test, (9.) obtained her regular license

Part II: Check students' work.

8.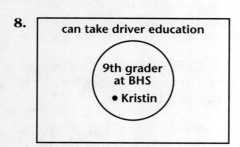

9. Kristin can take a driver education class.

10. The order of the statements should be: c, e, a, d, f, b.

ANSWERS

Long-Term Project — Chapter 3

1. multiples of 60°; 60°, 120°, 180°, 240°, 300°, 360°

2. 6

3. 720°

4. 120°

5. 90°

6. 25°; The diagonal bisects the angles.

7. 65°; △PIL has angles measuring 90, 25, and 65. ∠UTL and ∠TLI are congruent since they are alternate interior angles of the parallel sides.

8. No; rectangles have congruent diagonals, but rhombuses do not.

9. For Exercise 5, the angle would not be 90°. For Exercises 6 and 7, the angle measures would be different because of the angle formed by the diagonals. For Exercise 8, the diagonals would still not be congruent.

10. Answers may vary.

STATEMENTS	REASONS
$\overline{AB} \parallel \overline{CD}$ ∠ABC ≅ ∠BCD	Given If two lines cut by a transversal are parallel, then alternate interior angles are congruent.
$\overline{AB} \parallel \overline{EF}$ ∠ABC ≅ ∠BOF	Given If two lines cut by a transversal are parallel, then alternate interior angles are congruent.
$\overline{CD} \parallel \overline{GH}$ ∠BCD ≅ ∠BNH	Given If two lines cut by a transversal are parallel, then corresponding angles are congruent.
∠BOF ≅ ∠BNH	If two lines are cut by a transversal in such a way that corresponding angles are congruent, then the two lines are parallel.

$\overline{EF} \parallel \overline{GH}$ — If two lines cut by a transversal in such a way that corresponding angles are congruent, then the two lines are parallel.

11. 22

12. 97

13. 135°

14. 45°

15. 45°

16. 45°

17. 45°

18. 106

19. 106

20. 58°; ∠ADC is a right angle, so ∠ADE is 58°. △AED has angles of measure 58, 58, and 64.

21. The platform is about 82 by 150 feet. To estimate, you know that 150 feet is about 1.375 inches. The diagram is about 0.75 by 1.375 inches.

22. 108°

23. 108°

24. 36°

25. 36°

26. 36°

27. 72°

28. 72°

29. 36°

30. 72°

31. 72°

32. 108°

33. 108°

ANSWERS

34. Each interior angle of the pentagon is 108°. The triangles inside the pentagon are isosceles so base angles are congruent. Alternate interior angles are congruent.

35. 72°

36. 12°

37. 6 minutes

38. interior angles are 150°; exterior angles are 30°

39. 12 sides

40. $\frac{7}{3}$

41. -1

42. $\frac{2}{3}$

43. $-\frac{3}{4}$

44. 0

45. Answers may vary. $\overline{BC}$ is the most scary because the slope is down and the slope is greater than the other slopes.

46. undefined slope

47. Check students' work.

Part II: Check students' work.

Challenge: The area of ACTION GALORE! is about 9.5 acres. It is 2.5 acres larger than the park in Mall of America.

Long-Term Project — Chapter 4

Part I:

1. There are three sizes of squares. The medium square contains 4 congruent small squares. There is 1 octagon and 1 hexagon. There are 4 congruent isosceles right triangles and 4 smaller congruent scalene right triangles.

2. large square: 1 in., medium square: 0.5 in., small squares: 0.25 in., right isosceles triangle: legs of 0.5 in., other right triangles: legs of 0.25 and 0.5 in., octagon: sides of 1 in. and 0.625 in., hexagon: all sides about 0.5 in.

3. large square: 6 in., medium square: 3 in., small squares: 1.5 in., right isosceles triangle: legs of 3 in., other right triangles: legs of 1.5 and 3 in., octagon: sides of 6 in. and 3.75 in., hexagon: all sides about 3 in.

4. A 14–ft wide section has 14 squares. The 12–ft length has 12 · 14 = 168 sections.

5. Check students' work. There should be 4 tiles of the same size as the given diagram.

6. There are 4 congruent squares, 4 congruent rectangles, 4 congruent right triangles, 2 congruent isosceles triangles, and 2 congruent rhombuses. There is one square that encloses the rhombuses.

7. center square: 1.5 in., congruent squares: 0.25 in., congruent rectangles: 0.25 by 1.5 in., congruent right triangles: legs of 0.75 and 0.375 in., congruent isosceles triangles: base of 0.75 in. and legs of 0.875 in., rhombuses: about 0.875 in.

8. center square: 9 in., congruent squares: 1.5 in., congruent rectangles: 1.5 by 9 in., congruent right triangles: legs of 4.5 and 2.25 in., congruent isosceles triangles: base of 4.5 in. and legs of about 5.25 in., rhombuses: about 5.25 in.

9. You have 54 tiles. Answers may vary. 6 ft by 9 ft

10. Check students' work. There should be 4 tiles of the same size as the given diagram.

11. The design consists of all rectangles. Two pairs of rectangles are congruent.

ANSWERS

12. two congruent rectangles on the right side: 0.75 by 1 in.,
    top rectangle and congruent center rectangle: 1.25 by 0.25 in.,
    rectangle on left side: 1.75 by 0.25 in.,
    bottom rectangle: 0.5 by 1 in.,
    final rectangle: 0.75 by 1.25 in.

13. two congruent rectangles on the right side: 4.5 by 6 in.,
    top rectangle and congruent center rectangle: 7.5 by 1.5 in.,
    rectangle on left side: 10.5 by 1.5 in.,
    bottom rectangle: 3 by 6 in.,
    final rectangle: 4.5 by 7.5 in.

14. Check students' work. There should be 4 tiles of the same size as the given diagram.

15. There are 12 congruent rectangles, 11 congruent isosceles triangles, and 2 congruent right triangles.

16. 8 in. by 1.5 in.

17. There are 12 congruent rectangles, 18 congruent isosceles triangles, 2 other congruent isosceles triangles, and 8 congruent rhombuses.

18. top border — rectangles: 8 by 1.5 in., isosceles triangles: 8 in. base and 4.5 in. legs,
    right triangles: legs of 4 in. and 1.5 in.;
    center design — rectangles: 8 by 1.5 in., isosceles triangles: 5.3 in. base and 3 in. legs,
    end triangles: 3 in. legs and 4.5 in. base, rhombuses: 3 in. on a side

19. top border + top rows of squares + center design = 4.5 + 5(6) + 7.5 = 42 in., 8 ft − 42 in. = 54 in., 54 ÷ 6 = 9 rows

20. Check students' work. There should be a top border, 6 rows of 6 in. squares (8 across), a center design, and 9 rows of 6 in. squares.

Part II: Check students' work.

Long-Term Project — Chapter 5

Part I:

1. $4 \div 640 = 0.00625$

2. $1 \text{ mi}^2 = 5280^2 \text{ ft} = 27{,}878{,}400 \text{ ft}^2$; 4 acres $= 27{,}878{,}400 \cdot 0.00625 = 174{,}240 \text{ ft}^2$

3. $\sqrt{174240} \approx 417.42$; 417.42 ft by 417.42 ft

4. $4 \cdot 417.42 = 1669.68$ ft or about 1670 ft

5. An average walking speed is about 3 mph. That is 1 mi in 20 min. 1670 ft ≈ 0.32 or 1/3 mi; 1/3 mi would take $20 \div 3 \approx 6.7$ min.

6. Answers may vary.

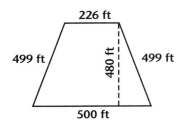

The trapezoid could have height 480 and bases of 226 and 500 ft. Using the Pythagorean Theorem and assuming the trapezoid is isosceles, the legs are 499 feet.

7. Answers may vary. $P = 226 + 2(499) + 500 = 1724$ ft

8.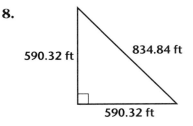

Since the triangle is an isosceles right triangle, the legs are equal. Then $A = 0.5bh$ or $0.5x^2$. The legs would be ≈ 590.32 ft and the hypotenuse would be ≈ $590.32\sqrt{2} \approx 834.84$ ft.

9. $P = 2(590.32) + 834.84 = 2015.48$ ft.

10. Answers may vary. The triangular lot has the greatest perimeter and would take the most time.

ANSWERS

11. $864 \div 640 = 1.35$ mi^2

12. Answers may vary. $1.35 = 0.5 \cdot 2.7$; 0.5 mi by 2.7 mi

13. Answers may vary. Using the answer to Exercise 12, $2(0.5) + 2(2.7) = 6.4$ mi.

14. Answers may vary. Using 3 mph, the time would be $6.4 \div 3 \approx 2.13$ hours or about 2 hours 8 min.

15.

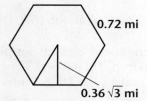

Let the hexagon side be s. Then the apothem is $0.5\sqrt{3}s$. $A = 0.5ap$, $1.35 = 0.5(0.5\sqrt{3}s)(6s)$, $s \approx 0.72$ mi and $a = 0.36\sqrt{3}s$ mi

16. $P = 6(0.72) = 4.32$ mi

17. Answers may vary.

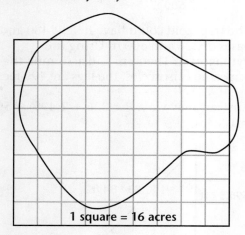

18. Answers may vary. Estimate the number of squares in the area and multiply by 16 acres since each square represents 16 acres. $16 \cdot 54 = 864$

19. Answers may vary. 16 acres = $16 \cdot 43,560 = 696,960$ ft^2; that gives about 834.84 ft on a side of a square. There are about 26 side lengths around the figure, so $P \approx 26 \cdot 834.84 = 21,705.84$ ft. Then $21,705.84 \div 5280 \approx 4.11$ mi.

20. Answers may vary. At a rate of 3 mph, the time would be about $4 \div 3 = 1.\overline{3}$ hr or 1 hr 20 min.

21. $150,000 \div 640 = 234.375$ mi^2

22. radius: $A = \pi r^2$, $234.375 = \pi r^2$, $r^2 \approx 74.6$, $r \approx 8.64$ mi; circumference: $C = 2\pi r$, $C = 2\pi(8.64) \approx 54.29$ mi

23. Answers may vary.

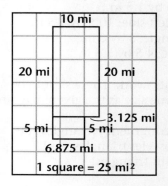

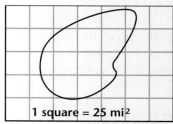

24. Answers may vary.

shape 1: $P = 20 + 10 + 20 + 5 + 6.875 + 5 + 3.125 = 70$ mi;
shape 2: squares are 5 mi on a side, so $P \approx 12 \cdot 5 = 60$ mi

25. Answers may vary. Use 60 mi as the perimeter. Use a driving speed of 25 mph since there's probably not a paved road. Then $60 \div 25 = 2.4$ hr or 2 hr 24 min.

Part II: Check students' work.

Long-Term Project — Chapter 6

The answers to many of these exercises can vary depending upon the accuracy of the students constructions. Accept answers that are reasonably close to the given answers.

ANSWERS

Part I:

1. about 200 mL

2. 40 + 40 + 20 + 20 + 50 + 50 = 220 cm²

3. 2 · 200 = 400 mL; Answers may vary.

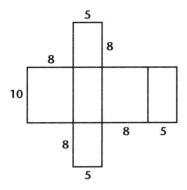

4. Answers may vary. 80 + 80 + 40 + 40 + 50 + 50 = 340 cm²

5. Answers may vary. $\frac{340}{220} = \frac{17}{11}$ ($\approx$ 1.55); The prism in Exercise 3 holds twice as much but the sum of the areas of the faces is not double. Students should express their own views on "twice as big."

6. 0.5 · 200 = 100 mL; Answers may vary.

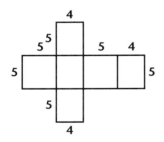

7. Answers may vary. 25 + 25 + 20 + 20 + 20 + 20 = 130 cm²

8. Answers may vary. $\frac{130}{220} = \frac{13}{22}$ ($\approx$ 0.59); The prism in Exercise 6 holds half as much but the sum of the areas of the faces is not half. Students should express their own views on "half as big."

9. about 400 mL

10. 88 + 88 + 88 + 52.4 + 52.4 $\approx$ 368.8 cm²

11. 2 · 400 = 800 mL; Answers may vary.

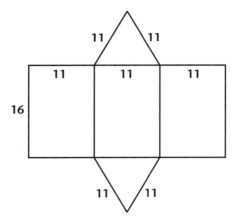

12. Answers may vary. 176 + 176 + 176 + 52.4 + 52.4 $\approx$ 632.8 cm²

13. Answers may vary. $\frac{632.8}{368.8} = \frac{79.1}{46.1}$ ($\approx$ 1.72); The prism in Exercise 11 holds twice as much but the sum of the areas of the faces is not double. Students should express their own views on "twice as big."

14. 0.5 · 400 = 200 mL; Answers may vary.

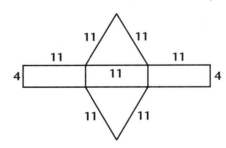

15. Answers may vary. 44 + 44 + 44 + 52.4 + 52.4 = 236.8 cm²

16. Answers may vary. $\frac{236.8}{368.8} = \frac{29.6}{46.1}$ ($\approx$ 0.64); The prism in Exercise 9 holds half as much but the sum of the areas of the faces is not half. Students should express their own views on "half as big."

ANSWERS

Part II: Check students' work. Approximate answers are given for Step 1.

Step 1:

1. about 512 mL

2. $64 + 64 + 64 + 64 + 64 + 64 = 384$ cm^2

3. $2 \cdot 512 = 1024$ mL; Students should draw a net for a cube that is about 10 cm on a side.

4. Answers may vary. $100 + 100 + 100 + 100 + 100 + 100 = 600$ cm^2

5. Answers may vary. $\frac{600}{384} = \frac{25}{16}$ $(= 1.5625)$; The prism in Exercise 3 holds twice as much but the sum of the areas of the faces is not double.

6. $0.5 \cdot 512 = 256$ mL; Students should draw a net for a cube that is about 6.3 cm on a side.

7. Answers may vary. $39.69 + 39.69 + + 39.69 + 39.69 + 39.69 + 39.69 = 238.14$ cm^2

8. Answers may vary. $\frac{238}{384} = \frac{119}{192}$ (≈ 0.62); The prism in Exercise 6 holds half as much, but the sum of the areas of the faces is not half. Students should express their own views on "half as big."

Long-Term Project — Chapter 7

Part I:

1. $r = 8$ ft $= 96$ in.; surface area $= \pi r^2 = \pi(96)^2 \approx 28,952.9$ in.^{2}; Answers may vary. The more exposed surface area, the more the water will cool down.

2. volume $= \pi r^2 h = \pi(96)^2(42) \approx 1,216,022.6$ in.$^3 \approx 5,264.2$ gallons

3. Use a formula to find the volume of the cylinder in cubic inches and then divide by 231 to get gallons.

4. 8,000 gal = 1,848,000 in.^{3}; Let x be the width of the rectangle. Then $V = \left[(x)(2x) + \pi\left(\frac{x}{2}\right)^2\right] \cdot h$; $V = 105.85x^2$; $105.85x^2 = 1,848,000$; $x \approx 132.13$ in.; dimensions of 8,000 gallon model are about 11 ft by 22 ft; 10,000 gal = 2,310,000 in.^{3}; Let x be the width of the rectangle. Then $V = \left[(x)(2x) + \pi\left(\frac{x}{2}\right)^2\right] \cdot h$; $V = 105.85x^2$; $105.85x^2 = 2,310,000$; $x \approx 147.73$ in.; dimensions of 10,000 gallon model are about 12.3 ft by 24.6 ft;

5. Answers may vary. An estimate is $18 \cdot 16 = 288$ ft^2. Count the full squares and combine partial squares to form more full squares. Multiply the number of full squares by 16.

6. 288 ft^2 = $288 \cdot 12 \cdot 12 = 41,472$ in.^{2}; $V = 41,472 \cdot 48 = 1,990,656$ in.^{3}; $1,990,656 \div 231 \approx 8,617.6$ gal

7.

8. $V = \left(\frac{1}{2}ap\right)h = \left(\frac{3\sqrt{3}}{2}\right)s^2(48)$; $10,000 = 2,310,000$ in.^{3}; $\left(\frac{3\sqrt{3}}{2}\right)s^2(48) = 2,310,000$; $124.71s^2 = 2,310,000$; $s \approx 136$ in.; 136 in. $= 11\frac{1}{3}$ ft

9. convert gallons to cubic inches, use the formula for the volume of a hexagonal prism, let s be the side length

ANSWERS

10.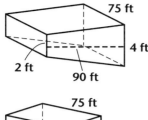
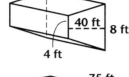
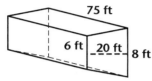

11. surface area = 150 · 75 = 11,250 ft^2

12. volume of prism 1 + volume of prism 2 + volume of prism 3 = $\frac{1}{2}$(24 + 48)(1080)(900) + (48 + 96)(480)(900) + $\frac{1}{2}$(72 + 96)(240)(900) = 34,992,000 + 31,104,000 + 18,144,000 = 84,240,000; 84,240,000 ÷ 231 ≈ 364,675.32 gallons

13.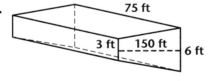
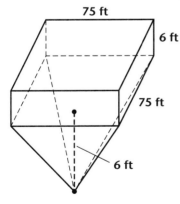

14. volume of trapezoidal prism + volume of rectangular prism + volume of pyramid = $\frac{1}{2}$(36 + 72)(1800)(900) + (900)(900)(72) + $\frac{1}{3}$(900)(900)(72) = 87,480,000 + 58,320,000 + 19,440,000 = 165,240,000 in.^{3}; 165,240,000 ÷ 231 ≈ 715,000 gallons

15. 1574 · 246 = 387,204 ft^2; 387,204 ÷ 43,560 ≈ 8.9 acres

16. (387,204 · 12^2 · 48) ÷ 231 ≈ 11,585,948 gallons

17. Answers may vary. Select 60 in. for the depth. $V = \pi r^2 \cdot 60$; (11,585,948) = 188.5r^2; $r ≈ 3768$ in. = 314 ft

Part II: Check students' work.

Long-Term Project — Chapter 8

Part I:

1. Check students' work.

2. Check students' work. Sample measurements are 162.5, 25, 160, 98, 38, 34, and 15.

3. Check students' work. The ratios for each measurement category should be very close.

4. In each column, the ratios are very close.

5–9. Answers will vary but should be close to the given approximations.

5. 1.0

6. 0.14

7. 0.62

8. 0.23

9. 0.47

ANSWERS

10. 8 ft 11 in. = 107.1 in.;
wing span: $\frac{1}{1} = \frac{x}{107.1}$, $x = 107.1$ in.;
foot length: $\frac{14}{100} = \frac{x}{107.1}$, $x \approx 15$ in.;
navel: $\frac{62}{100} = \frac{x}{107.1}$, $x \approx 66.4$ in.;
tibia: $\frac{23}{100} = \frac{x}{107.1}$, $x \approx 24.6$ in.

11. set up a proportion such as $\frac{14}{100} = \frac{x}{107.1}$, where 0.14 is the ratio of foot length, and solve for x

12. Answers may vary. Set up a proportion using my neck to height ratio. $\frac{34}{162.5} = \frac{x}{107.1}$, $x \approx 22.4$ in. Then use the ratio 0.47 for wrist to neck and set up a proportion to find his wrist using the calculated measurement for his neck. $\frac{47}{100} = \frac{x}{22.4}$, $x \approx 10.5$ in.

13. wing span: $\frac{1}{1} = \frac{x}{24}$, $x = 24$ in.;
foot length: $\frac{14}{100} = \frac{x}{24}$, $x \approx 3.4$ in.;
navel: $\frac{62}{100} = \frac{x}{24}$, $x \approx 14.9$ in.;
tibia: $\frac{23}{100} = \frac{x}{24}$, $x \approx 5.5$ in.

14. set up a proportion such as $\frac{14}{100} = \frac{x}{24}$, where 0.14 is the ratio of foot length, and solve for x

15. Answers may vary. Set up a proportion using my neck to height ratio. $\frac{34}{162.5} = \frac{x}{24}$, $x \approx 5.0$ in. Then use the ratio 0.47 for wrist to neck and set up a proportion to find her wrist using the calculated measurement for her neck. $\frac{47}{100} = \frac{x}{5.0}$, $x \approx 2.4$ in.

16. The unit of measure has no effect on the ratios. The ratios for centimeters will be the same as the ratio for feet and/or inches.

17. Answers may vary. Using a collector's doll, $\frac{36}{162.5} \approx 0.22$

18. Answers may vary.
foot length: $25 \cdot 0.22 = 5.5$ cm;
navel: $98 \cdot 0.22 = 21.6$ cm;
tibia: $38 \cdot 0.22 = 8.4$ cm;
wrist: $15 \cdot 0.22 = 3.3$ cm;
neck: $34 \cdot 0.22 = 7.5$ cm

19. Answers may vary. Multiplied the ratio of doll height: my height times each of my measurements to get the doll's measurements.

20. Answers may vary. The measurements were fairly close. The doll is similar to me.

21. Answers may vary. Count the whole squares and combine partial squares to get more whole squares. The area was about 150 cm².

22. Answers may vary. If the scale factor of lengths is 0.22, then the area scale will be $(0.22)^2 = 0.0484$. Set up and solve the proportion $\frac{484}{10000} = \frac{x}{150}$, $x \approx 7.26$ cm.

23. Answers may vary. The area of the actual doll's foot was about 7 cm², very close to the estimate.

Part II: Check students' work.

Long-Term Project — Chapter 9

Part I:

1. 10 cm square, 10 cm

2. $\frac{8}{9} \cdot 10 = \frac{80}{9}$; $\left(\frac{80}{9}\right)^2 = \frac{6400}{81} \approx 79$ in.²

3. $A = 5\pi r^2 = \pi(5)^2 \approx 78.5$ in.²; The answers are very close.

ANSWERS

4. Answers may vary.
   $d = 15$ cm, $A = \left(\frac{8}{9}d\right)^2 = \left(\frac{8}{9} \cdot 15\right)^2 \approx 177.8$ in.^{2}, $A = \pi r^2 = \pi(7.5)^2 \approx 176.7$ in.^{2};
   $d = 40$ cm, $A = \left(\frac{8}{9}d\right)^2 = \left(\frac{8}{9} \cdot 40\right)^2 \approx 1264.2$ in.^{2}, $A = \pi r^2 = \pi(20)^2 \approx 1256.6$ in.^{2}
   The answers are very close.

5. Check students' drawings. The length should be about 3.125 in.

6. The length of the segment is very close to the value of π.

7. Lengths should be close to the following: for the circle of radius 2 in., the length is about 6.25 in. $\approx 2\pi$; for the circle of radius 3 in., the length is about 9.375 in. $\approx 3\pi$

8. $\pi \approx 4 \cdot \left(\frac{2}{3} + \frac{2}{35} + \frac{2}{99}\right) \approx 2.976$

9. $\pi \approx 4 \cdot \left(1 - \frac{1}{3} + \frac{1}{5} - \frac{1}{7} + \frac{1}{9} - \frac{1}{11} + \frac{1}{13} - \frac{1}{15} + \frac{1}{17} - \frac{1}{19} + \frac{1}{21} - \frac{1}{23}\right)$
   $\approx 4 \cdot \left(\frac{2}{3} + \frac{2}{35} + \frac{2}{99} + \frac{2}{195} + \frac{2}{325} + \frac{2}{483}\right)$
   ≈ 3.0584

10. Check students' tables.

11–12 (and Part II Steps 1 & 2) Answers may vary for columns 2, 3, and 5, but the students' answers in columns 4 and 6 should be very close to those shown in this table.

Col. 1	Col. 2	Col. 3	Col. 4	Col. 5	Col. 6
tri	1.19	6.21	2.609	12.42	5.218
squ	1.4	7.91	2.825	11.19	3.996
pent	1.67	9.77	2.925	12.08	3.617
hex	1.25	7.5	3.000	8.66	3.464
hept	1.51	9.2	3.046	10.21	3.381
oct	1.22	7.47	3.061	8.09	3.316
non	1.34	8.26	3.082	8.78	3.276
deca	1.18	7.26	3.076	7.63	3.233
11	1.4	8.7	3.107	9.07	3.239
dodec	2.3	14.28	3.104	14.78	3.213

13. Draw segment $\overline{OA}$. Construct a perpendicular to $\overline{OA}$ at A since a tangent to a circle is perpendicular to a radius at the point of tangency.

14. For $\triangle ABC$: By construction, $\angle BOA = 120°$. Then $\widehat{AB} = 120°$. By the Inscribed Angle Theorem, $\angle BCA = \frac{1}{2}\widehat{AB} = \frac{1}{2}(120°) = 60°$. This same reasoning can be used for the other two angles of $\triangle ABC$. For $\triangle DEF$: By construction, $\angle BOA = 120°$. Then $\widehat{AB} = 120°$. The measure of a tangent–tangent angle is one–half the difference of the intercepted arcs. So, $\angle EFD = \frac{1}{2}(240° - 120°) = 60°$. This same reasoning can be used for the other two angles of $\triangle DEF$.

Part II: Check students' work. Look at the table for Exercises 11 and 12.

Long-Term Project — Chapter 10

Part I:
1. $\triangle AOB = 36°$, $\angle OBA = 90°$, $\angle OAB = 54°$

2. 2 in.

3. $\tan \angle OAB = \frac{OB}{2}$; $OB = 2 \tan 54°$

4. 20 in.

5. $A = \frac{1}{2}ap = \frac{1}{2}(2 \tan 54°)(20) = 20 \tan 54° \approx 27.53$ in.^{2}

6. $p = 10x$; $a = x \tan 54°$;
   $A = \frac{1}{2}ap = \frac{1}{2}(x \tan 54°)(10x)$
   $= 5x^2 \tan 54°$

7. for $\triangle DEH$: $\angle EDH = 108°$, $\angle DEH = 36°$; $\angle DHE = 36°$;
   for $\triangle HEG$: $\angle HEG = 36°$, $\angle EHG = 72°$; $\angle EGH = 72°$;
   for $\triangle FEG$: $\angle EFG = 108°$, $\angle FEG = 36°$; $\angle FGE = 36°$

8. $\triangle DEH \cong \triangle FEG$; All angles and all side lengths are congruent.

ANSWERS

9. $HE^2 = y^2 + y^2 - 2(y)(y)\cos 108°$;
$HE = \sqrt{2y^2 - 2y^2\cos 108°}$
$= \sqrt{2y^2(1 - \cos 108°)}$
$= y\sqrt{2(1 - \cos 108°)}$
$\approx 1.618y$

10. $p = \frac{1}{2}(y + y + 1618y) = 1.809y$

11. $A = \sqrt{1.809y(1809y - y)(1.809y - y)(1.809y - 1.618y)}$
$= \sqrt{1.809y(0.809y)(0.809y)(0.191y)}$
$= \sqrt{0.226y^4}$
$= 0.475y^2$; so $A = 0.475y^2$

12. $p = \frac{1}{2}(y + 1.618y + 1.618y) = 2.118y$

13. $A = \sqrt{2.118y(2.118y - y)(2.118y - 1.618y)(2.118y - 1.618y)}$
$= \sqrt{2.118y(1.118y)(0.5y)(0.5y)}$
$= \sqrt{0.592y^4}$
$= 0.769y^2$; or $A = 0.769y^2$

14. $A = 0.475y^2 + 0.769y^2 + 0.475y^2$
$= 1.719y^2$; or $A = 1.719y^2$

15. $A = 1.719y^2 = 1.719(4)^2 \approx 27.504$ in.^{2}

16. $A = \frac{1}{2}ap = \frac{1}{2}(2.75)(20) = 27.5$ in.^{2}
The answers are very close.

17. $\angle AOE = 72°$, $\angle OEA = \angle OAE = 54°$

18. $\angle AFO = 90°$, $\angle FOA = 36°$, $\angle OAF = 54°$

19. Answers may vary. Sample using the cosine of 54°: $\cos 54° = \frac{AF}{4}$; $AF = 4\cos 54°$

20. $p = (10)4\cos 54° = 40\cos 54° \approx 23.5$ in.

21. $p = 10z \cos 54°$

22. Answers may vary. $4\frac{7}{8} \cdot 5 = 24\frac{3}{8} = 24.375$;
The answers are very close.

Part II: Check students' work. Possible formulas are given in the table.

Polygon	Area #1 ($x = 0.5$ · side length)	Area #2 ($y =$ side length)	Perimeter ($z =$ radius of circumscribed circle)
hexagon	$6x^2 \tan 60°$	$2.598y^2$	$12z \cos 60°$
octagon	$8x^2 \tan 67.5°$	$4.828y^2$	$16z \cos 67.5°$
nonagon	$9x^2 \tan 70°$	$6.182y^2$	$18z \cos 70°$
decagon	$10x^2 \tan 72°$	$7.694y^2$	$20z \cos 72°$
dodecagon	$12x^2 \tan 75°$	$11.196y^2$	$24z \cos 75°$

ANSWERS

Long-Term Project — Chapter 11

Part I:

1. Check students' tables. A spreadsheet can be used.

2. Check students' work.

3. 1.6; golden ratio

4. Check that students have a box-like structure.

5. Check that students have the first box and then two smaller boxes at this step.

6. The fractal has one box that is 4 in. by 2.5 in. and 2 boxes that are 2 in. by 1.25 in.

7. The fractal has one box that is 4 in. by 2.5 in., 2 boxes that are 2 in. by 1.25 in., and 4 boxes that are 1 in by 0.625 in.

8. The fractal has one box that is 4 in. by 2.5 in., 2 boxes that are 2 in. by 1.25 in., 4 boxes that are 1 in by 0.625 in., and 8 boxes that are 0.5 in. by 0.3125 in.

9.

#	Placement of cut	Length of cut	Dimensions of inner rectangle	Ratio of l to w
1	2	2.5	4 × 2.5	1.6
2	1	1.25	2 × 1.25	1.6
3	0.5	0.625	1 × 0.625	1.6
4	0.25	0.3125	0.5 × 0.3125	1.6
5	0.125	0.15625	0.25 × 0.15626	1.6
6	0.0625	0.078125	0.125 × 0.078125	1.6
7	0.03125	0.0390625	0.0625 × 0.0390625	1.6
8	0.015625	0.0195312	0.03125 × 0.0195312	1.6
9	0.0078125	0.0097656	0.015626 × 0.0097656	1.6
10	0.0039062	0.0048828	0.0078125 × 0.0048828	1.6

patterns: column 2 and 3 — each entry is one-half the previous entry; column 4 — all entries have a length to width ratio of 1.6

Geometry

ANSWERS

10. A spreadsheet can be used.

Stage #	# of Boxes Added	Total # of Boxes
1	1	1
2	2	3
3	4	7
4	8	15
5	16	31
6	32	63
7	64	127
8	128	255
9	256	511
10	512	1023

11. If n is the stage number, then the number of boxes added is 2^{n-1}. The total number of boxes at any stage is $2^n - 1$.

12. Check students' tables. A spreadsheet can be used.

13. 20 in.^{2}

14. 1

15. 5 in.^{2}

16. 10 in.^{2}; It is double the area.

17. 5 in.^{2}, from Exercise 15

18. 1.25

19. 1.25 in.^{2}

20. 1.3125

21. 1.328125

22. The total surface area at any stage divided by the surface area at Stage 1 approaches 1.33. ...

Stage #	New Surface Area Added	Total Surface Area	Total SA ÷ SA at Stage 1
1	20	20	1
2	5	25	1.25
3	1.25	26.25	1.3125
4	0.3125	26.5625	1.328125
5	0.078125	26.640625	1.33203125
6	0.01953125	26.66015625	1.333007813
7	0.004882813	26.66503906	1.333251953
8	0.001220703	26.66625977	1.333312988
9	0.000305176	26.66656494	1.333328247
10	7.62939E-05	26.66664124	1.333332062

23 e. It is like a Sierpinski gasket. The outer shape is a triangle and then there are pop–up boxes added at each iteration.

Part II: Check students' work.

ANSWERS

Long-Term Project — Chapter 12

1. size and color

2. Check students' work.

4. Answers may vary. (Check students' Euler diagrams for their 4 cases.) the set of small squares is a subset of the set of all squares, the set of blue triangles is a subset of the set of triangles; the set of blue triangles is a subset of the set of blue figures, the set of circles is a subset of the set of circles

5. Answers may vary. the set of real numbers, the set of integers

6. Answers may vary. set of 14-year-olds, set of 16-year-olds, set of 19-year-olds

7. No, he may be 14, 15, 16, 17, 18, or 19 since those are also subsets of the set of teenagers.

8. a. valid
   b. invalid; Ian could be outside of the circle for getting a seat
   c. valid
   d. invalid; Idaho could be outside of the circle for being a state
   e. valid

9. Check students' work.

10. red circles should be in both sets since a red circle is both red and a circle

11. red small circle and red large circle

12. Answers may vary. The figure is blue and a square. The figure is yellow and large. The figure is small and a triangle. The figure is large and a circle.

13. 12

14. Answers may vary. The figure is blue or a square or a blue square. The figure is yellow or large or a large, yellow figure. The figure is small or a triangle or a small triangle. The figure is large or a circle or a large circle.

Part II: Check students' work.